SECOND EDITION

How to Pass

NATIONAL 5

Maths

Robert Barclay

HODDER
GIBSON
AN HACHETTE UK COMPANY

The Publishers would like to thank the following for permission to reproduce copyright material:

The table on pages 101–104 has been adapted from the National 4 Mathematics Course Support Notes and is reproduced with the kind permission of SQA – © Scottish Qualifications Authority.

Orders: please contact Bookpoint Ltd, 130 Park Drive, Milton Park, Abingdon, Oxon OX14 4SE. Telephone: (44) 01235 827827. Fax: (44) 01235 400454. Lines are open 9.00–5.00, Monday to Saturday, with a 24-hour message answering service. Visit our website at www.hoddereducation.co.uk. Hodder Gibson can also be contacted directly at hoddergibson@hodder.co.uk

© Robert Barclay 2018

First published in 2018 by
Hodder Gibson, an imprint of Hodder Education,
An Hachette UK Company
211 St Vincent Street
Glasgow G2 5QY
Impression number: 2

Year: 2019

Cover photo © Maxal Tamor/Shutterstock
Illustrations by Aptara, Inc.
Typeset in Cronos Pro 13/15 by Aptara, Inc.
Printed in India
A catalogue record for this title is available from the British Library
ISBN: 978 1 5104 2099 1

Contents

Welcome to *How to Pass National 5 Maths: Second Edition*

I am glad that you are working towards passing National 5 Mathematics this session and that you have chosen this book to help you achieve your full potential.

Ever since you started school, and even before that, you have been building up your mathematical knowledge and skills. Now the time has come to practise everything you have learned and need to know in order to do well in your National 5 Mathematics course assessment.

This book is designed to help give you every chance of achieving the best possible grade.

In this book you will find:
- information about National 5 Mathematics
  - an outline of the course and how it is assessed
  - comprehensive details of the course content, including worked examples with explanations
- information about the maths you are expected to have experienced prior to embarking on the National 5 course
  - a summary of the National 4 Mathematics course content
- advice on
  - working throughout the year
  - revising effectively
  - improving your mark.

The National 5 Mathematics course

This course is suitable for learners who have achieved the fourth level of learning across the mathematics experiences and outcomes in the broad general education, or who have attained the National 4 Mathematics course, or who have equivalent qualifications or experience. An overview of the National 4 Mathematics course is provided in Appendix 2.

During the course you will extend your knowledge, understanding and skills in algebra, geometry, trigonometry, numeracy, statistics and reasoning.

Throughout the course you will be expected to use reasoning skills to interpret information, to select a strategy to solve problems and to communicate solutions.

The course content is summarised in the following table. Further details relating to each topic are provided in Chapters 1 to 18 of this book.

Algebra	**Geometry**	**Trigonometry**
<ul><li>Expanding brackets</li><li>Factorising</li><li>Completing the square</li><li>Algebraic fractions</li><li>Equation of straight line</li><li>Functional notation</li><li>Equations and inequations</li><li>Simultaneous equations</li><li>Change of subject of formulae</li><li>Graphs of quadratic functions</li><li>Quadratic equations</li></ul>	<ul><li>Gradient</li><li>Arc and sector of circle</li><li>Volume</li><li>Pythagoras' Theorem</li><li>Properties of shapes</li><li>Similarity</li><li>Vectors and 3D coordinates</li></ul>	<ul><li>Graphs</li><li>Equations</li><li>Identities</li><li>Area of triangle, sine rule, cosine rule, bearings</li></ul>
	Numeracy	**Statistics**
	<ul><li>Surds</li><li>Indices and scientific notation</li><li>Rounding (significant figures)</li><li>Percentages</li><li>Fractions</li></ul>	<ul><li>Semi-interquartile range, standard deviation</li><li>Scattergraph; equation of line of best fit</li></ul>

Reasoning
- Interpreting a situation where mathematics can be used and identifying a strategy.
- Explaining a solution and/or relating it to context.

National 5 Mathematics course assessment

The course assessment is an examination comprising two question papers.

The number of marks and times allotted for the papers are as follows.

Paper 1 (non-calculator)	50 marks	1 hour 15 minutes
Paper 2	60 marks	1 hour 50 minutes

The course assessment is graded A–D; the grade is determined by the total mark you score in the examination.

Formulae list

You are given the formulae listed below in the examination.

You are expected to know all other formulae that you may need to use.

The roots of $ax^2 + bx + c = 0$ are given by $x = \dfrac{-b \pm \sqrt{(b^2 - 4ac)}}{2a}$

Sine rule: $\dfrac{a}{\sin A} = \dfrac{b}{\sin B} = \dfrac{c}{\sin C}$

Cosine rule: $a^2 = b^2 + c^2 - 2bc \cos A$ or $\cos A = \dfrac{b^2 + c^2 - a^2}{2bc}$

Area of a triangle: $\text{Area} = \frac{1}{2} ab \sin C$

Volume of a sphere: $\text{Volume} = \frac{4}{3} \pi r^3$

Volume of a pyramid: $\text{Volume} = \frac{1}{3} Ah$

Volume of a cone: $\text{Volume} = \frac{1}{3} \pi r^2 h$

Standard deviation $s = \sqrt{\dfrac{\sum(x - \bar{x})^2}{n-1}} = \sqrt{\dfrac{\sum x^2 - \dfrac{(\sum x)^2}{n}}{n-1}}$, where n is the sample size.

Work throughout the year

The best way to succeed in maths is to practise, practise, practise!

Work hard throughout the year. It is the cumulative effect of your daily efforts that will help you to do well in the examination. Always complete unfinished class work at home and do all homework that is set. Many questions in textbooks and in homework exercises are similar to examination questions, so you gain valuable practice from doing these. In maths, what you learn one day often forms the basis of what you learn the next. So, in class, don't hesitate to ask about things you don't understand. Attend all revision classes that are on offer.

Practise using **your own** calculator throughout the year. Some calculators work differently, so make sure that you can use yours properly to carry out all the different types of calculations you might meet in the examination. This will increase your chance of doing well in Paper 2.

Revise effectively

Make **effective** use of study time. Keep all your books and equipment together so that you don't waste time at the start of each study session looking for your revision materials.

Make sure you have:
- How to Pass National 5 Maths book
- past papers or specimen papers
- class notes, summaries and jotters
- paper, pencils, ruler, rubber, protractor and a set of compasses
- scientific calculator.

Study in a quiet place free from potential distractions. Do not study with the television on, or beside others who may distract you. Switch your mobile telephone off.

Studying for short periods, often and regularly is the recipe for success. You should take a 5–10 minute break every hour to allow your brain to relax. You concentrate much better if you study for short periods but often.

Doing maths questions is the most valuable use of your study time. You will benefit much more spending 30 minutes doing maths questions than spending several hours copying out notes or reading a maths textbook. The more questions you do, the more you will get correct, the more confident you will become and the better you will do in the examination.

Practise doing the type of questions that are likely to appear in the examination. Work through past papers or, in the case of a new examination, specimen papers. Use the marking instructions to check your answers and to understand what the examiners are looking for.

Improving your mark

- The most important piece of advice is **SHOW ALL YOUR WORKING CLEARLY** in every question.
  The instructions in the examination paper state that 'Full credit will only be given where the solution contains appropriate working.'
  A 'correct' answer with no working may only be awarded partial marks or even no marks at all.
  An incomplete answer will be awarded marks for any appropriate working.
  Attempt every question, even if you are not sure whether you are correct or not. Your solution may contain working which will gain some marks. A blank response is certain to be awarded no marks. Never cross out working unless you have something better to replace it with.

- The examination is designed so that, for most people, the questions get slightly harder as you work through the paper. So attempt the questions in the order that they occur to ensure that you gain the marks that are likely to be easier to achieve before tackling the harder questions.
- Do not leave the examination room before the end of the allotted time. Spend any time you have left to check the answers you have given. If there are any questions you have been unable to answer, you may remember how to do them in the last few minutes of the examination. The extra marks you gain might mean the difference between passing and failing, or between an A grade and a B grade.
- Where a question has more than one part, but you are unable to answer the first part, still attempt subsequent parts. You may still gain full marks in the subsequent parts.

For example, consider this question.

A rectangular lawn is $(x + 2)$ metres long and $(x - 1)$ metres wide. The area of the lawn is 40 square metres.

$(x + 2)$ m

$(x - 1)$ m

a) Show that $x^2 + x - 42 = 0$.
b) Find the length of the lawn.

Even if you are unable to do part (a), you can still score full marks in part (b) for correctly solving the equation $x^2 + x - 42 = 0$ and stating the length of the lawn.

Chapter 1

Significant figures, percentages and fractions

What you should know

You should know how to:
* ★ round to a given number of significant figures
* ★ use percentages to calculate compound interest, appreciation and depreciation
* ★ use reverse percentages to calculate an original quantity
* ★ add, subtract, multiply and divide common fractions including mixed numbers (and work with combinations of these operations).

Rounding

Answers to calculations are often rounded to an appropriate degree of accuracy. For example, answers to money calculations are often rounded to 2 decimal places (to the nearest penny). For example, 17·5% of £29·99 = £5·24825 = £5·25 to 2 decimal places. A number rounded to n **decimal places**, has n digits to the right of the decimal point.

Another kind of rounding uses **significant figures**. All figures in a number are significant except:
* zeros at the **end** of a whole number
* zeros at the **start** of a number with a decimal point.

Example

1 70 518 has 5 significant figures.
70 518 = 70 520 correct to 4 sig. figs. ——— Zeros at the **end** of a whole number are not significant.
70 518 = 70 500 correct to 3 sig. figs.
70 518 = 71 000 correct to 2 s.f.
70 518 = 70 000 correct to 1 s.f.

2 0·003907 has 4 significant figures
0·003907 = 0·00391 correct to 3 sig. figs. ——— Zeros at the **start** of a number with a decimal point are not significant.
0·003907 = 0·0039 correct to 2 s.f.
0·003907 = 0·004 correct to 1 s.f.

Compound interest

Interest is money paid to you by a bank or building society for saving with them (or charged to you for borrowing from them).

The original amount you save (or borrow) is called the principal.

Compound interest is interest paid on both the principal and any interest already added.

Example

Morag invests £8000 in a savings account for 3 years at an interest rate of 4% per annum.

Calculate the amount in Morag's account at the end of the 3 years.

Solution

Each year the amount is 104% (100% + 4%) of the previous year's amount.

After 1 year: amount = 1.04×8000 = £8320 —— Since $104\% = \frac{104}{100} = 1.04$

After 2 years: amount = 1.04×8320 = £8652·80

or $(1.04)^2 \times 8000$ = £8652·80

After 3 years: amount = $(1.04)^3 \times 8000$ = £8998·91 (to the nearest penny)

Appreciation

When the value of something increases we say its value has appreciated.

Example

A house valued at £180 000 at the end of 2011 appreciated in value by 5% in 2012 and by 4·5% in 2013. Calculate its value at the end of 2013.

Solution

After 1 year the house was worth 105% (100% + 5%) of its value at the end of 2011.

After 2 years the house was worth 104·5% (100% + 4·5%) of its value at the end of 2012.

End of 2012:   value = $1.05 \times 180\,000$ = £189 000 —— Since $105\% = \frac{105}{100} = 1.05$

End of 2013: value = $1.045 \times 189\,000$ = £197 505 —— Since $104.5\% = \frac{104.5}{100} = 1.045$

Depreciation

When the value of something decreases we say its value has depreciated.

Example

A car bought for £15 000 depreciated by 20% during its first year and by a further 15% during its second year. Calculate its value after two years.

Solution

After 1 year the car was worth 80% (100% − 20%) of its original value.

After 2 years the car was worth 85% (100% − 15%) of its value after 1 year.

After 1 year: value = $0.8 \times 15\,000$ = £12 000 —— Since $80\% = \frac{80}{100} = \frac{8}{10} = 0.8$

After 2 years: value = $0.85 \times 12\,000$ = £10 200 —— Since $85\% = \frac{85}{100} = 0.85$

Reversing a percentage change

1 The price of a family room for two nights at the Bay Hotel is £210 including Value Added Tax (VAT) at 20%.

Calculate the price excluding VAT.

Solution

Price including VAT   = 120% (100% + 20%) of price excluding VAT.

This means   £210 = 1·2 × price excluding VAT.

So price excluding VAT = £210 ÷ 1·2

                           = £175.

2 There are 720 students in Westside High School this year. This is 4% less than last year.

How many students were in Westside High School last year?

Solution

Number of students this year = 96% (100% − 4%) of number of students last year.

This means   720 = 0·96 × number of students last year.

So number of students last year = 720 ÷ 0·96

                                   = 750.

Fractions

You will be expected to carry out calculations like these in the non-calculator paper.

1 Calculate $\frac{1}{2} + \frac{3}{7}$.

Solution

$\frac{1}{2} + \frac{3}{7} = \frac{7}{14} + \frac{6}{14}$ ——— Make the denominators the same.

$= \frac{13}{14}$ ——— Add the numerators.

Do not add the denominators.

2 Calculate $\frac{4}{5} - \frac{7}{15}$.

Solution

$\frac{4}{5} - \frac{7}{15} = \frac{12}{15} - \frac{7}{15}$ ——— Make the denominators the same.

$= \frac{\cancel{5}}{\cancel{15}^{3}}$ ——— Subtract the numerators.

$= \frac{1}{3}$ ——— Divide both numerator and denominator by 5.

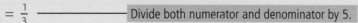

⇨

3 Calculate $1\frac{4}{9} + 2\frac{5}{6}$.

Solution

$1\frac{4}{9} + 2\frac{5}{6} = 1\frac{8}{18} + 2\frac{15}{18}$ —— Make the denominators the same.

$\quad = 3\frac{23}{18}$ —————— Add the whole numbers and add the numerators.

$\quad = 4\frac{5}{18}$ —————— Convert $\frac{23}{18}$ to $1\frac{5}{18}$.

4 Calculate $6\frac{1}{4} - 4\frac{2}{3}$.

Solution

$6\frac{1}{4} - 4\frac{2}{3} = 6\frac{3}{12} - 4\frac{8}{12}$ —— Make the denominators the same.

$\quad = 5\frac{15}{12} - 4\frac{8}{12}$ —— Convert $6\frac{3}{12}$ to $5 + 1\frac{3}{12} = 5\frac{15}{12}$

$\quad = 1\frac{7}{12}$ ———— Subtract the whole numbers and subtract the numerators.

5 Calculate $\frac{4}{9} \times \frac{15}{16}$.

Solution

$\frac{\overset{1}{\cancel{4}}}{9} \times \frac{15}{\underset{4}{\cancel{16}}} = \frac{1}{\underset{3}{\cancel{9}}} \times \frac{\overset{5}{\cancel{15}}}{4}$ —— Divide both the 4 and the 16 by 4.

$\quad = \frac{1}{3} \times \frac{5}{4}$ ———— Divide both the 9 and the 15 by 3.

$\quad = \frac{5}{12}$ ————— Multiply numerators together and multiply denominators together.

6 Calculate $2\frac{5}{8} \times 3\frac{1}{3}$.

Solution

$2\frac{5}{8} \times 3\frac{1}{3} = \frac{\overset{7}{\cancel{21}}}{8} \times \frac{10}{\underset{1}{\cancel{3}}}$ —— Change the mixed numbers to vulgar fractions.

$\quad = \frac{7}{\underset{4}{\cancel{8}}} \times \frac{\overset{5}{\cancel{10}}}{1}$ —— Divide both the 21 and the 3 by 3.

$\quad = \frac{7}{4} \times \frac{5}{1}$ ———— Divide both the 8 and the 10 by 2.

$\quad = \frac{35}{4}$ ————— Multiply numerators together and multiply denominators together.

$\quad = 8\frac{3}{4}$ ————— Change the vulgar fraction to a mixed number.

7 Calculate $\frac{7}{12} \div \frac{5}{8}$.

Solution

$\frac{7}{12} \div \frac{5}{8} = \frac{7}{\underset{3}{\cancel{12}}} \times \frac{\overset{2}{\cancel{8}}}{5}$ —— Change the ÷ to a × and turn the second fraction upside down.

$\quad = \frac{7}{3} \times \frac{2}{5}$ ———— Divide both the 12 and the 8 by 4.

$\quad = \frac{14}{15}$ ————— Multiply numerators together and multiply denominators together.

⇨

8 Calculate $2\frac{2}{3} \div 1\frac{1}{5}$.

Solution

$2\frac{2}{3} \div 1\frac{1}{5} = \frac{8}{3} \div \frac{6}{5}$ —————— Change the mixed numbers to vulgar fractions.

$= \frac{\overset{4}{\cancel{8}}}{3} \times \frac{5}{\underset{3}{\cancel{6}}}$ —————— Change the ÷ to a × and turn the second fraction upside down.

$= \frac{4}{3} \times \frac{5}{3}$ —————— Divide both the 8 and the 6 by 2.

$= \frac{20}{9}$ —————— Multiply numerators together and multiply denominators together.

$= 2\frac{2}{9}$ —————— Change the vulgar fraction to a mixed number.

9 Calculate $2\frac{1}{5} + \frac{7}{10}$ of $1\frac{6}{7}$.

Solution

$\frac{7}{10}$ of $1\frac{6}{7} = \frac{7}{10} \times 1\frac{6}{7}$ —————— Calculate $\frac{7}{10}$ of $1\frac{6}{7}$ first. Remember BODMAS.

$= \frac{\overset{1}{\cancel{7}}}{10} \times \frac{13}{\underset{1}{\cancel{7}}}$ —————— Convert the mixed number to a vulgar fraction.

$= \frac{1}{10} \times \frac{13}{1}$ —————— Divide both the 7s by 7.

$= \frac{13}{10}$ —————— Multiply numerators together and multiply denominators together.

$= 1\frac{3}{10}$ —————— Convert the vulgar fraction to a mixed number.

—————— Now add the answer to $2\frac{1}{5}$.

$2\frac{1}{5} + 1\frac{3}{10} = 2\frac{2}{10} + 1\frac{3}{10}$ —————— Make the denominators the same.

$= 3\frac{\overset{1}{\cancel{5}}}{\underset{2}{\cancel{10}}}$ —————— Add the whole numbers and add the numerators.

$= 3\frac{1}{2}$ —————— Divide both numerator and denominator by 5.

10 Calculate $\frac{4}{5}$ of $(1\frac{8}{9} - \frac{1}{2})$.

Solution

$1\frac{8}{9} - \frac{1}{2} = 1\frac{16}{18} - \frac{9}{18}$ —————— Calculate $1\frac{8}{9} - \frac{1}{2}$ first. Remember BODMAS.

—————— Make the denominators the same.

$= 1\frac{7}{18}$ —————— Subtract the numerators.

$\frac{4}{5}$ of $1\frac{7}{18} = \frac{4}{5} \times 1\frac{7}{18}$ —————— Now calculate $\frac{4}{5}$ of the answer.

$= \frac{\overset{2}{\cancel{4}}}{5} \times \frac{25}{\underset{9}{\cancel{18}}}$ —————— Convert the mixed number to a vulgar fraction.

$= \frac{2}{\underset{1}{\cancel{5}}} \times \frac{\overset{5}{\cancel{25}}}{9}$ —————— Divide both the 4 and the 18 by 2.

$= \frac{2}{1} \times \frac{5}{9}$ —————— Divide both the 5 and the 25 by 5.

$= \frac{10}{9}$ —————— Multiply the numerators together and multiply the denominators together.

$= 1\frac{1}{9}$ —————— Convert the vulgar fraction to a mixed number.

For practice

(The answers to the following questions are given in Appendix 1.)

1 Round
- **a)** 52 390 to 1 significant figure
- **b)** 0·0666 to 2 s.f.
- **c)** 208 517 to 3 s.f.

2 Ruth invested £2400 in a savings account for 3 years.

The interest rate was 4% per annum in the first year, 3·5% per annum in the second year and 3·75% per annum in the third year.

Calculate how much Ruth had in the account after the 3 years.

3 A cottage bought for £80 000 three years ago appreciated in value by 5% per annum.

How much is the cottage worth now?

4 Jim buys a motor bike for £3000. Its value depreciates by 25% over the first year and by a further 20% during the second year.

How much is Jim's motor bike worth after the two years?

5 After receiving a 6% pay rise, Julie is now paid £18 550 per year.

How much was Julie paid per year before the pay rise?

6 A store reduces all its prices by 35% in a sale. The sale price of a laptop is £299.

Calculate the price of the laptop before the sale.

7 Calculate $\frac{2}{3} + \frac{4}{7}$.

8 Calculate $2\frac{3}{4} + 3\frac{1}{5}$.

9 Calculate $\frac{5}{6} - \frac{3}{8}$.

10 Calculate $6\frac{1}{2} - 1\frac{9}{10}$.

11 Calculate $\frac{7}{12} \times \frac{8}{9}$.

12 Calculate $1\frac{7}{8} \times 2\frac{2}{9}$.

13 Calculate $\frac{3}{4} \div \frac{5}{6}$.

14 Calculate $1\frac{3}{5} \div 4\frac{2}{3}$.

15 Calculate $3\frac{3}{4} - \frac{2}{3}$ of $1\frac{5}{8}$.

16 Calculate $\frac{3}{7}$ of $(2\frac{1}{6} + \frac{3}{4})$.

You should know how to:
★ work with algebraic expressions involving expansion of brackets
★ factorise an algebraic expression
★ complete the square in a quadratic expression.

Expanding brackets and collecting like terms

To remove brackets from an expression, multiply each term in the bracket by the term outside the bracket, then collect like terms.

Example

1 $5(x + 2) + 2(3x - 7)$
$= 5x + 10 + 6x - 14$
$= 11x - 4$

2 $3a(4a + 3)$
$= 12a^2 + 9a$

3 $1 - 4(2p - 5)$
$= 1 - 8p + 20$
$= 21 - 8p$

4 $(x + 5)(x + 3)$
$= x(x + 3) + 5(x + 3)$
$= x^2 + 3x + 5x + 15$
$= x^2 + 8x + 15$

5 $(3n - 2)(4n + 1)$
$= 3n(4n + 1) - 2(4n + 1)$
$= 12n^2 + 3n - 8n - 2$
$= 12n^2 - 5n - 2$

6 $(y + 2)(y^2 - 3y + 4)$
$= y(y^2 - 3y + 4) + 2(y^2 - 3y + 4)$
$= y^3 - 3y^2 + 4y + 2y^2 - 6y + 8$
$= y^3 - y^2 - 2y + 8$

Hints & tips

When squaring out a pair of brackets, it saves time if you apply these two results.

$$(x + y)^2 = (x + y)(x + y)$$
$$= x(x + y) + y(x + y)$$
$$= x^2 + xy + yx + y^2$$
$$= x^2 + 2xy + y^2$$

e.g. $(t + 6)^2 = t^2 + 2 \times t \times 6 + 6^2$
$$= t^2 + 12t + 36$$

$$(x - y)^2 = (x - y)(x - y)$$
$$= x(x - y) - y(x - y)$$
$$= x^2 - xy - yx + y^2$$
$$= x^2 - 2xy + y^2$$

e.g. $(3c - 5)^2 = (3c)^2 - 2 \times 3c \times 5 + 5^2$
$$= 9c^2 - 30c + 25$$

Factorising

Factorising is the inverse process of expanding brackets.

Factorising using a common factor

Example

1 $8a + 12b$
$$= 4(2a + 3b)$$

2 $x^2 - 7x$
$$= x(x - 7)$$

3 $3m^2 + 15mn$
$$= 3m(m + 5n)$$

Remember

The highest common factor must be taken outside the bracket to complete the factorisation.

Factorising differences of two squares

Example

1 $x^2 - y^2$
$$= (x + y)(x - y)$$

2 $p^2 - 36$
$$= (p + 6)(p - 6)$$

3 $9c^2 - 16d^2$
$$= (3c + 4d)(3c - 4d)$$

Factorising trinomials

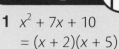

1 $x^2 + 7x + 10$
$= (x + 2)(x + 5)$

2 $a^2 - 2a - 15$
$= (a + 3)(a - 5)$

3 $2n^2 - 7n + 6$
$= (2n - 3)(n - 2)$

Hints & tips

You can check that you have factorised an expression correctly by multiplying out the brackets.

Further factorising

Factorise in the following order:

- take out any common factor
- factorise any difference of two squares
- factorise any trinomial that remains.

Example

1 $3x^2 - 75$
$= 3(x^2 - 25)$
$= 3(x + 5)(x - 5)$

2 $2s^2 + 6s - 8$
$= 2(s^2 + 3s - 4)$
$= 2(s + 4)(s - 1)$

Completing the square in a quadratic expression

The process of converting the expression $x^2 + bx + c$ into the form $(x + p)^2 + q$ is known as completing the square.

Example

1 $x^2 + 6x - 1$
$\frac{1}{2}$ of 6 is 3 and
$(x + 3)^2 = x^2 + 6x + 9$
So $x^2 + 6x - 1 = x^2 + 6x + 9 - 10$
$= (x + 3)^2 - 10.$

2 $y^2 - 10y + 27$
$\frac{1}{2}$ of -10 is -5 and
$(y - 5)^2 = y^2 - 10y + 25$
So $y^2 - 10y + 27 = y^2 - 10y + 25 + 2$
$= (y - 5)^2 + 2.$

Remember

To write $x^2 + bx + c$ in completed square form:
write the **square term** as $(x + \frac{1}{2}b)^2$ and then adjust for the **constant value.**

For practice ⦾

(The answers to the following questions are given in Appendix 1.)

1 Expand the brackets:
   a) $2x(x - 5)$
   b) $-y(3y - 1)$

2 Expand the brackets and simplify:
   a) $19 - 7(a + 2)$
   b) $4(b - 2) + 3(5b + 1)$
   c) $6(2c + 3) - 2(c - 7)$
   d) $(x + 9)(x - 4)$
   e) $(2y + 9)(5y + 3)$
   f) $(n - 8)^2$
   g) $(2k + 7)^2$
   h) $(m + 3)(m^2 + 2m - 5)$
   i) $(2t - 1)(4t^2 - 3t + 5)$

3 Factorise:
   a) $24x - 18y$
   b) $35ab + 5b$
   c) $2c^2 + 10c$
   d) $m^2 - 81$
   e) $25 - n^2$
   f) $9p^2 - 100q^2$
   g) $x^2 - 3x - 10$
   h) $y^2 - 9y + 8$
   i) $3s^2 + 11s + 6$

4 Factorise fully:
   a) $2x^2 - 18$
   b) $5d^2 + 10d - 40$
   c) $6k^2 + 15k + 9$

5 Express in completed square form i.e. in the form $(x + p)^2 + q$:
   a) $x^2 + 8x + 10$
   b) $x^2 - 14x + 50$
   c) $x^2 + 2x - 1$
   d) $x^2 - 4x - 5$
   e) $x^2 + 3x + 2$

Indices, scientific notation and surds

What you should know

You should know how to:
* ★ simplify expressions using the laws of indices
* ★ carry out calculations using scientific notation
* ★ simplify surds and carry out calculations with surds
* ★ rationalise denominators.

Indices

The term 2^5 (read as '2 to the power 5') is a short way of writing $2 \times 2 \times 2 \times 2 \times 2$.

The 5 is called the **index** or the power. The plural of index is indices.

The 2 is called the **base**.

Numbers and expressions written in the form a^n are in index form.

Note that $\boldsymbol{a^1 = a}$.

Example

Law 1

   (i) $2^3 \times 2^5 = 2^8$

   (ii) $a^4 \times a^7 = a^{11}$

Law 2

   (i) $7^6 \div 7^2 = 7^4$

   (ii) $\frac{b^9}{b^4} = b^5$

   (iii) $c^8 \div c^7 = c^1 = c$

Law 3

   (i) $(4^3)^5 = 4^{15}$

   (ii) $(d^2)^6 = d^{12}$

Law 4

$(2e^4)^3 = 2^3 \times (e^4)^3 = 8e^{12}$

Remember

The laws of indices
1. $a^m \times a^n = a^{m+n}$
2. $a^m \div a^n = a^{m-n}$
3. $(a^m)^n = a^{mn}$
4. $(ab)^n = a^n b^n$

Key points

The laws of indices do not apply to terms with **different bases**.

For example $7^2 \times 2^4 \neq (7 \times 2)^{2+4}$. This is because

$(7 \times 2)^{2+4} = 14^6 = 7\ 529\ 536$.

The correct calculation is $7^2 \times 2^4 = 49 \times 16 = 784$.

Example

1 (i) $13^0 = 1$
   (ii) $b^0 = 1$
   (iii) $c^6 \div c^6 = c^0 = 1$
2 (i) $3^{-2} = \frac{1}{3^2} = \frac{1}{9}$

   (ii) $d^{-4} = \frac{1}{d^4}$
   (iii) $e^{-3} \times e^{-5} = e^{-8} = \frac{1}{e^8}$
3 (i) $8^{\frac{2}{3}} = \sqrt[3]{8^2} = 2^2 = 4$
   (ii) $f^{\frac{7}{4}} = \sqrt[4]{f^7}$
   (iii) $\left(g^{\frac{2}{5}}\right)^3 = g^{\frac{6}{5}} = \sqrt[5]{g^6}$

> **Remember**
>
> Zero, negative and fractional indices
> 1 $a^0 = 1$
> 2 $a^{-n} = \frac{1}{a^n}$
> 3 $a^{m/n} = \sqrt[n]{a^m}$

Key points

* The square root of x can be written as $x^{\frac{1}{2}}$, since $\sqrt{x} = \sqrt[2]{x^1} = x^{\frac{1}{2}}$.

* The cube root of x can be written as $x^{\frac{1}{3}}$, since $\sqrt[3]{x} = \sqrt[3]{x^1} = x^{\frac{1}{3}}$.

* The fourth root of x can be written as $x^{\frac{1}{4}}$, since $\sqrt[4]{x} = \sqrt[4]{x^1} = x^{\frac{1}{4}}$.

Example

1 $49^{\frac{1}{2}} = \sqrt{49} = 7$
2 $1000^{\frac{1}{3}} = \sqrt[3]{1000} = 10$ —— Since $10^3 = 10 \times 10 \times 10 = 1000$
3 $16^{\frac{1}{4}} = \sqrt[4]{16} = 2$ —— Since $2^4 = 2 \times 2 \times 2 \times 2 = 16$

Example

Simplify:

1 $\dfrac{4a^{-7} \times 3a^4}{2a^{-5}}$

2 $\dfrac{18b^{\frac{2}{3}}}{6b^{\frac{1}{2}}}$

Solution

1 $\dfrac{4a^{-7} \times 3a^4}{2a^{-5}} = \dfrac{12a^{-3}}{2a^{-5}}$ —— Law 1 – add indices.

   $= 6a^2$ —— Law 2 – subtract indices.

2 $\dfrac{18b^{\frac{2}{3}}}{6b^{\frac{1}{2}}} = \dfrac{18b^{\frac{4}{6}}}{6b^{\frac{3}{6}}}$

   $= 3b^{\frac{1}{6}}$ —— Law 2 – subtract indices.

$\Rightarrow$

➡️

3 $3c^{-1}(c^3 - 2c)$

4 $(d^{\frac{1}{2}} + d^{-\frac{1}{2}})(d^{\frac{3}{2}} + d^{-\frac{1}{2}})$

Solution

3 $3c^{-1}(c^3 - 2c) = 3c^2 - 6c^0$ — Expand brackets using Law 1; remember that $c = c^1$.

$\qquad\qquad\quad = 3c^2 - 6$ — Remember that $c^0 = 1$.

4 $(d^{\frac{1}{2}} + d^{-\frac{1}{2}})(d^{\frac{3}{2}} + d^{-\frac{1}{2}})$

$\quad d^{\frac{1}{2}}(d^{\frac{3}{2}} + d^{-\frac{1}{2}}) + d^{-\frac{1}{2}}(d^{\frac{3}{2}} + d^{-\frac{1}{2}})$

$\quad = d^2 + d^0 + d^1 + d^{-1}$ —— Expand brackets using Law 1.

$\quad = d^2 + 1 + d + \frac{1}{d^1}$ —— Remember that $d^0 = 1$ and $d^{-1} = \frac{1}{d^1}$.

$\quad = d^2 + 1 + d + \frac{1}{d}$ —— Remember that $d^1 = d$.

Scientific notation

Scientific notation or standard form is a shorthand way of writing very large and very small numbers.

A number written in scientific notation is written in the form $a \times 10^n$ where $1 \le a < 10$ and n is an integer.

Key points

The number $149\,000\,000 = 1{\cdot}49 \times 100\,000\,000$
$\qquad\qquad\qquad\qquad = 1{\cdot}49 \times 10^8.$

So the number $149\,000\,000$ can be written in scientific notation as $1{\cdot}49 \times 10^8$.

The rules for doing this are:
* to find a, put the decimal point between the 1 and the 4 (i.e. $a = 1{\cdot}49$)
* to find n, count the number of places you must move the decimal point to the **right** to get $149\,000\,000$ (i.e. $n = 8$)
* so $149\,000\,000 = 1{\cdot}49 \times 10^8$.

A **large** number has a **positive power** of 10.

The number $0{\cdot}000\,037 = 3{\cdot}7 \times 0{\cdot}000\,01$
$\qquad\qquad\qquad\qquad = 3{\cdot}7 \times 10^{-5}.$

So the number $0{\cdot}000\,037$ can be written in scientific notation as $3{\cdot}7 \times 10^{-5}$.

The rules for doing this are:
* to find a, put the decimal point between the 3 and the 7. $a = 3{\cdot}7$
* to find n, count the number of places you must move the decimal point to the **left** to get $0{\cdot}000\,037$. Thus $n = -5$ (*n is negative when the decimal point is moved left*).
* so $0{\cdot}000\,037 = 3{\cdot}7 \times 10^{-5}$.

A **small** number has a **negative power** of 10.

Calculations involving scientific notation

When **multiplying** numbers in scientific notation (without a calculator) remember to **add** the powers of ten. When **dividing, subtract** the powers of ten.

> ## Example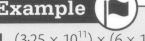
>
> **1** $(3{\cdot}25 \times 10^{11}) \times (6 \times 10^{-2})$
> $= (3{\cdot}25 \times 6) \times (10^{11} \times 10^{-2})$
> $= 19{\cdot}5 \times 10^{9}$
> $= 1{\cdot}95 \times 10^{10}$
>
> **2** $(1{\cdot}8 \times 10^{6}) \div (2 \times 10^{13})$
> $= (1{\cdot}8 \div 2) \times (10^{6} \div 10^{13})$
> $= 0{\cdot}9 \times 10^{-7}$
> $= 9 \times 10^{-8}$

Remember

Remember how to write numbers in scientific notation correctly. For example, 149 000 000 in scientific notation is $1{\cdot}49 \times 10^{8}$, not $1{\cdot}49^{8}$ or 149×10^{6}.

To do calculations such as those above on your calculator you need to know how to use the [EXP] or [EE] or [×10ˣ] button.

> ## Example
>
> **1** $(3{\cdot}25 \times 10^{11}) \times (6 \times 10^{-2})$
> Enter: 3·25[EXP]11 × 6[EXP](−2) =
> This will display: 1·95 EXP 10
> So the answer is: $1{\cdot}95 \times 10^{10}$
>
> **2** $(1{\cdot}8 \times 10^{6}) \div (2 \times 10^{13})$
> Enter: 1·8[EXP]6 ÷ 2[EXP]13 =
> This will display: 9EXP −8
> So the answer is: 9×10^{-8}

Surds

In maths you have worked with various **sets** of numbers.

Natural numbers $N = \{1, 2, 3, \ldots\}$

Whole numbers $W = \{0, 1, 2, 3, \ldots\}$

Integers $Z = \{\ldots -2, 1, 0, 1, 2, \ldots\}$

Rational numbers $Q = \{$numbers which can be written as a ratio, $\dfrac{a}{b}$, of two integers$\}$

Some examples of rational numbers are:

a) $2 \left(= \frac{2}{1}\right)$

b) $0{\cdot}7 \left(= \frac{7}{10}\right)$

c) $1\frac{2}{3}\left(= \frac{5}{3}\right)$.

Numbers which cannot be written as a ratio of two integers are called **irrational** numbers. The number $\pi = 3\cdot14159\ldots$ is an example of an irrational number.

Roots of rational numbers which cannot be expressed as rational numbers are called **surds**. A surd is an irrational number. These are examples of surds: $\sqrt{2}$, $\sqrt{3}$, $\sqrt{0.4}$, $\sqrt[3]{10}$.

Numbers like $\sqrt{49}$, $\sqrt{0.16}$, $\sqrt[3]{8}$ are not surds because they are rational numbers.

Simplifying surds

Key points

1 $\sqrt{ab} = \sqrt{a} \times \sqrt{b}$ —— To simplify a surd, express it as a product of two numbers, one of which is a perfect square.

2 $\sqrt{\dfrac{a}{b}} = \dfrac{\sqrt{a}}{\sqrt{b}}$

Example

Express in simplest form:

a) $\sqrt{24}$

b) $\sqrt{80}$

c) $\sqrt{\dfrac{4}{25}}$

d) $\sqrt{\dfrac{27}{300}}$

Solution

a) $\sqrt{24} = \sqrt{4} \times \sqrt{6}$
$\qquad = 2\sqrt{6}$

b) $\sqrt{80} = \sqrt{16} \times \sqrt{5}$ —— Choose the factor pair of 80 with the **highest** perfect square.
$\qquad = 4\sqrt{5}$

c) $\sqrt{\dfrac{4}{25}} = \dfrac{\sqrt{4}}{\sqrt{25}}$
$\qquad = \dfrac{2}{5}$

d) $\sqrt{\dfrac{27}{300}} = \dfrac{\sqrt{27}}{\sqrt{300}}$
$\qquad = \dfrac{\sqrt{9} \times \sqrt{3}}{\sqrt{100} \times \sqrt{3}}$
$\qquad = \dfrac{3\sqrt{3}}{10\sqrt{3}}$
$\qquad = \dfrac{3}{10}$

Calculations involving surds

The normal rules of algebra for collecting like terms and multiplying out brackets apply to calculations involving surds.

Express in simplest form:

a) $5\sqrt{7} - 3\sqrt{7}$

b) $\sqrt{45} + \sqrt{125}$

c) $\sqrt{3}(2\sqrt{6} - \sqrt{11})$

d) $(\sqrt{6} + \sqrt{2})^2$

Solution

a) $5\sqrt{7} - 3\sqrt{7} = 2\sqrt{7}$

b) $\sqrt{45} + \sqrt{125} = \sqrt{9} \times \sqrt{5} + \sqrt{25} \times \sqrt{5}$

$\qquad\qquad\qquad = 3\sqrt{5} + 5\sqrt{5}$

$\qquad\qquad\qquad = 8\sqrt{5}$

c) $\sqrt{3}(2\sqrt{6} - \sqrt{11}) = 2\sqrt{18} - \sqrt{33}$

$\qquad\qquad\qquad\quad = 2 \times \sqrt{9} \times \sqrt{2} - \sqrt{33}$

$\qquad\qquad\qquad\quad = 2 \times 3 \times \sqrt{2} - \sqrt{33}$

$\qquad\qquad\qquad\quad = 6\sqrt{2} - \sqrt{33}$

d) $(\sqrt{6} + \sqrt{2})^2 = (\sqrt{6})^2 + 2 \times \sqrt{6} \times \sqrt{2} + (\sqrt{2})^2$

$\qquad\qquad\qquad = 6 + 2\sqrt{12} + 2$

$\qquad\qquad\qquad = 8 + 2 \times \sqrt{4} \times \sqrt{3}$

$\qquad\qquad\qquad = 8 + 2 \times 2 \times \sqrt{3}$

$\qquad\qquad\qquad = 8 + 4\sqrt{3}$

Rationalising the denominator

When the denominator of a fraction is a surd we usually remove the surd from the denominator. This process is called rationalising the denominator.

Rationalise the denominator and simplify where possible:

a) $\frac{2}{\sqrt{3}}$

b) $\frac{5}{2\sqrt{10}}$

Solution

a) $\frac{2}{\sqrt{3}} = \frac{2}{\sqrt{3}} \times \frac{\sqrt{3}}{\sqrt{3}} = \frac{2\sqrt{3}}{3}$ ⎯⎯ Multiplying by $\frac{\sqrt{3}}{\sqrt{3}}$ does not change the value of the original number, since $\frac{\sqrt{3}}{\sqrt{3}} = 1$.

b) $\frac{5}{2\sqrt{10}} = \frac{5}{2\sqrt{10}} \times \frac{\sqrt{10}}{\sqrt{10}} = \frac{5\sqrt{10}}{20} = \frac{\sqrt{10}}{4}$

Example

Rationalise the denominator of $\frac{1}{\sqrt{5}+1}$.

Solution

In this case multiply by $\frac{\sqrt{5}-1}{\sqrt{5}-1}$. The expression $\sqrt{5}-1$ is called the **conjugate** of $\sqrt{5}+1$.

So $\frac{1}{\sqrt{5}+1} \times \frac{\sqrt{5}-1}{\sqrt{5}-1} = \frac{\sqrt{5}-1}{(\sqrt{5}+1)(\sqrt{5}-1)} = \frac{\sqrt{5}-1}{5-\sqrt{5}+\sqrt{5}-1} = \frac{\sqrt{5}-1}{4}$.

For practice

(The answers to the following questions are given in Appendix 1.)

1 Evaluate (giving your answer as a fraction): $2^0 - 6^{-1}$

2 Evaluate:

  a) $27^{\frac{2}{3}}$ b) $25^{\frac{3}{2}}$ c) $16^{\frac{1}{4}}$

3 Simplify:

  a) $\frac{3a^5 \times 4a^{-1}}{6a}$

  b) $b^8 \times (b^3)^{-2}$

  c) $\frac{c^2}{\sqrt{c}}$

  d) $(2d^3)^4 \div d^{-2}$

  e) $(5e^4)^2 \times 2e$

4 Simplify and give your answer with a positive index: $\frac{a^{\frac{1}{4}} \times a^{-\frac{5}{4}}}{a^3}$

5 Expand the brackets and simplify:

  a) $m^{\frac{1}{2}}(m^{\frac{1}{2}} + 2m^{-\frac{1}{2}})$ b) $(n^{\frac{4}{3}} + n^{\frac{1}{3}})(n^{\frac{2}{3}} - n^{-\frac{1}{3}})$

6 Write the following numbers in full.

  a) The population of India is $1{\cdot}23 \times 10^9$.

  b) The diameter of a human hair is $2{\cdot}1 \times 10^{-4}$ mm.

7 Write these numbers in scientific notation.

  a) The population of the United Kingdom is 63 million.

  b) The mass of a grain of sand is 0·008 grams.

8 Calculate:

  a) $(4{\cdot}36 \times 10^{-3}) \times (2{\cdot}5 \times 10^{12})$

  b) $(2{\cdot}432 \times 10^{-4}) \div (6{\cdot}4 \times 10^9)$

9 The speed of light is 3×10^8 metres per second. The distance from the Sun to Earth is $1{\cdot}5 \times 10^{11}$ metres. How long does it take for light to travel from the Sun to Earth?

10 Given that $T = 3n^2$, find the value of T when $n = 8 \times 10^7$.

11 Express in simplest form:

  a) $\sqrt{18}$ b) $\sqrt{48}$ c) $\sqrt{\frac{36}{49}}$ d) $\sqrt{\frac{12}{75}}$

12 Express in simplest form:

  a) $\sqrt{27} + \sqrt{75} - \sqrt{3}$ b) $\sqrt{2}(\sqrt{10} + 3\sqrt{2})$ c) $(\sqrt{5}+2)(\sqrt{5}-2)$ d) $(\sqrt{3}-\sqrt{2})^2$

13 Rationalise the denominator and simplify where possible:

  a) $\frac{3}{\sqrt{5}}$ b) $\frac{1}{2\sqrt{2}}$ c) $\frac{9}{\sqrt{6}}$ d) $\frac{1}{(3-\sqrt{2})}$

Algebraic fractions

Reducing algebraic fractions to simplest form

This is known as simplifying a fraction.

Key points

To write a fraction in its simplest form:
* ✻ factorise the numerator and denominator
* ✻ divide the numerator and denominator by their highest common factor. This step is called cancelling a fraction.

Remember

$$\frac{24}{30} = \frac{6 \times 4}{6 \times 5} = \frac{4}{5}$$

Example

1. $\frac{3}{12a} = \frac{\cancel{3}^{1} \times 1}{\cancel{3}_{1} \times 4 \times a} = \frac{1}{4a}$

2. $\frac{7m}{mn} = \frac{7 \times \cancel{m}^{1}}{\cancel{m}_{1} \times n} = \frac{7}{n}$

3. $\frac{6c}{9c^2} = \frac{\cancel{3}^{1} \times 2 \times \cancel{c}^{1}}{\cancel{3}_{1} \times 3 \times \cancel{c}_{1} \times c} = \frac{2}{3c}$

4. $\frac{2x - 8}{(x - 4)^2} = \frac{2\cancel{(x - 4)}^{1}}{\cancel{(x - 4)}_{1}(x - 4)} = \frac{2}{x - 4}$

5. $\frac{x^2 + 2x - 15}{x^2 - 25} = \frac{\cancel{(x + 5)}^{1}(x - 3)}{\cancel{(x + 5)}_{1}(x - 5)} = \frac{x - 3}{x - 5}$

Adding and subtracting algebraic fractions

Key points

To add or subtract fractions:
* ✻ convert the original fractions to equivalent fractions with the same denominator; often the easiest way to find common denominator is to multiply the original denominators
* ✻ add or subtract the numerators; do **not** add or subtract the denominators
* ✻ check that your answer is in simplest form.

Remember

$$\frac{1}{3} + \frac{5}{8} = \frac{8}{24} + \frac{15}{24} = \frac{23}{24}$$

Example

1 $\frac{1}{a} + \frac{9}{b} = \frac{b}{ab} + \frac{9a}{ab} = \frac{b + 9a}{ab}$

2 $\frac{x + 1}{5} - \frac{x - 2}{3} = \frac{3(x + 1)}{15} - \frac{5(x - 2)}{15}$

$= \frac{3(x + 1) - 5(x - 2)}{15}$

$= \frac{3x + 3 - 5x + 10}{15}$

$= \frac{13 - 2x}{15}$

3 $\frac{5}{x + 1} - \frac{3}{x - 2} = \frac{5(x - 2)}{(x + 1)(x - 2)} - \frac{3(x + 1)}{(x + 1)(x - 2)}$

$= \frac{5(x - 2) - 3(x + 1)}{(x + 1)(x - 2)}$

$= \frac{5x - 10 - 3x - 3}{(x + 1)(x - 2)}$

$= \frac{2x - 13}{(x + 1)(x - 2)}$

4 $\frac{2}{n} + \frac{n - 1}{n^2} = \frac{2n}{n^2} + \frac{n - 1}{n^2} = \frac{2n + n - 1}{n^2} = \frac{3n - 1}{n^2}$

Multiplying and dividing algebraic fractions

Key points

To multiply fractions:

* ❋ multiply the numerators, then multiply the denominators
* ❋ express the fraction in simplest form by cancelling where necessary.

To divide fractions:

* ❋ multiply the first fraction by the inverse of the second fraction.

Remember

a) $\frac{3}{4} \times \frac{6}{7} = \frac{18}{28} = \frac{9}{14}$

b) $\frac{5}{9} \div \frac{2}{3} = \frac{5}{9} \times \frac{3}{2} = \frac{15}{18} = \frac{5}{6}$

Example

1 $\frac{3a}{8} \times \frac{2b}{a} = \frac{\overset{3}{\cancel{6}} \, \overset{1}{\cancel{a}} \, b}{\underset{4}{\cancel{8}} \, \underset{1}{\cancel{a}}} = \frac{3b}{4}$

2 $\frac{5m}{6n} \div \frac{4m^2}{3n} = \frac{5m}{6n} \times \frac{3n}{4m^2} = \frac{\overset{5}{\cancel{15}} \, \overset{1}{\cancel{m}} \, \overset{1}{\cancel{n}}}{\underset{8}{\cancel{24}} \, \underset{m}{\cancel{m^2}} \, \underset{1}{\cancel{n}}} = \frac{5}{8m}$

For practice

(The answers to the following questions are given in Appendix 1.)

1 Simplify:

a) $\frac{4ab}{6ab^2}$

b) $\frac{x - 2}{5x - 10}$

c) $\frac{n^2 + 3n}{n^2 - 9}$

d) $\frac{x^2 - 5x + 4}{x^2 + 2x - 3}$

e) $\frac{2x^2 + 5x + 3}{(x + 1)^2}$

2 Express each of the following as single fractions in simplest form.

a) $\frac{2}{b} + \frac{3}{4c}$

b) $\frac{3}{y} - \frac{y}{3}$

c) $\frac{x - 1}{2} + \frac{x + 4}{5}$

d) $\frac{3}{n - 5} - \frac{2}{n - 2}$

e) $\frac{5}{2x} + \frac{1}{x^2}$

f) $\frac{c}{d^2} \times \frac{7d}{2c}$

g) $\frac{4u^2}{15v^2} \div \frac{8u}{9v}$

Equations, inequalities and change of subject

What you should know

You should know how to:
* ★ solve linear equations and inequalities
* ★ change the subject of a formula.

Solving linear equations

To solve an equation:
* add or subtract the same amount to or from each side
* multiply or divide each side by the same number.

Example

Solve for x:

1 $\frac{x}{2} + 5 = 12$

$\Rightarrow \frac{x}{2} = 7$ —— Undo the +5 by subtracting 5 from each side of the equation.

$\Rightarrow x = 14$ —— Undo the ÷2 by multiplying each side of the equation by 2.

2 $4x - 1 = x + 6$

$\Rightarrow 3x - 1 = 6$ —— Undo the +x by subtracting x from each side of the equation.

$\Rightarrow 3x = 7$ —— Undo the −1 by adding 1 to each side of the equation.

$\Rightarrow x = \frac{7}{3}$ —— Undo the ×3 by dividing each side of the equation by 3.

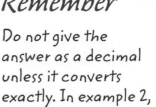

Remember

Do not give the answer as a decimal unless it converts exactly. In example 2, $\frac{7}{3} = 2\cdot333\ldots$, so the answer is $x = \frac{7}{3}$, not $x = 2\cdot33$.

Equations with brackets

Example

Solve for x:

1 $7x - 2(x - 6) = 9$

$\Rightarrow 7x - 2x + 12 = 9$

$\Rightarrow 5x + 12 = 9$

$\Rightarrow 5x = -3$

$\Rightarrow x = -\frac{3}{5}$

2 $(x + 1)^2 = (x - 1)(x + 5)$

$\Rightarrow x^2 + 2x + 1 = x^2 + 4x - 5$

$\Rightarrow 2x + 1 = 4x - 5$

$\Rightarrow -2x + 1 = -5$

$\Rightarrow -2x = -6$

$\Rightarrow -x = -3$

$\Rightarrow x = 3$

Inequalities

An inequality is a mathematical statement such as $x > 3$, $y \le 8$ etc.

The same rules for solving equations can be applied to inequalities, with one exception. When you multiply or divide both sides of an inequality by a negative number the inequality is reversed.

Example

1 $9x + 5 > 3$
$\Rightarrow 9x > -2$
$\Rightarrow x > \frac{-2}{9}$

2 $2x - 1 \le 6x + 11$
$\Rightarrow -4x - 1 \le 11$
$\Rightarrow -4x \le 12$
$\Rightarrow x \ge -3$

Alternative method: $2x - 1 \le 6x + 11$
$\Rightarrow -1 \le 4x + 11$
$\Rightarrow -12 \le 4x$
$\Rightarrow -3 \le x$
$\Rightarrow x \ge -3$

Remember

$x > 3$ means that x is greater than 3
$y \ge 3$ means that y is greater than or equal to 3
$x < 8$ means that x is less than 8
$y \le 8$ means that y is less than or equal to 8.

Changing the subject of a formula

A formula is an equation which shows the relationship between different quantities.

For example, the formula $C = \frac{5}{9}(F - 32)$ can be used to change temperatures in degrees Fahrenheit to degrees Celsius. Here C is called the **subject** of the formula. If we wanted to change a temperature in degrees Celsius to degrees Fahrenheit it would help if the formula was rearranged to make F the subject.

Example

1 Make x the subject of the formula $y = 5x - 2$.
2 Make r the subject of the formula $A = \frac{\pi r^2}{3}$.
3 Make b the subject of the formula $L = 3a - \sqrt{b}$.

Solution

1 $y = 5x - 2 \Rightarrow 5x - 2 = y$
$\Rightarrow 5x = y + 2$
$\Rightarrow x = \frac{y + 2}{5}$

2 $A = \frac{\pi r^2}{3} \Rightarrow \frac{\pi r^2}{3} = A$
$\Rightarrow \pi r^2 = 3A$
$\Rightarrow r^2 = \frac{3A}{\pi}$
$\Rightarrow r = \sqrt{\frac{3A}{\pi}}$

3 $L = 3a - \sqrt{b} \Rightarrow L + \sqrt{b} = 3a$
$\Rightarrow \sqrt{b} = 3a - L$
$\Rightarrow b = (3a - L)^2$

For practice

(The answers to the following questions are given in Appendix 1.)

1 Solve:
   a) $11x + 8 = 9x - 6$
   b) $1 - 4x = 3$
   c) $\frac{x}{4} - \frac{1}{2} = 5$
   d) $\frac{3x + 2}{5} = 7$
   e) $x - 4(x + 1) = 6$
   f) $x(x + 2) = (x + 6)(x - 1)$

2 Solve:
   a) $3x + 2 \geq 5x - 6$
   b) $9 - x < 3(x + 4)$

3 Change the subject of the formula $k = 4m - 3n$ to m.
4 Change the subject of the formula $P = 2(L + B)$ to B.
5 Change the subject of the formula $V = \frac{1}{3}Ah$ to h.
6 Change the subject of the formula $a = \frac{x}{b} - c$ to x.
7 Change the subject of the formula $T = uv^2 + 2$ to v.
8 Change the subject of the formula $T = \frac{kn^2}{5}$ to n.
9 Change the subject of the formula $L = p + \sqrt{q}$ to q.

The equation of a straight line

What you should know

You should know how to:
* ★ determine the gradient of a straight line
* ★ determine the equation of a straight line
* ★ use functional notation.

Gradient

Gradient of a line = $\dfrac{\text{vertical height}}{\text{horizontal distance}}$

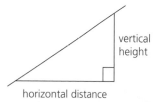

Example

gradient = $\frac{4}{1}$ = 4

gradient = $\frac{-2}{3}$

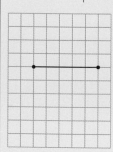

gradient = $\frac{0}{5}$ = 0

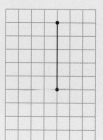

gradient = $\frac{5}{0}$ i.e. undefined

Remember

☞ Line slopes up from left to right ⇒ positive gradient.

☞ Line slopes down from left to right ⇒ negative gradient.

☞ Horizontal line ⇒ gradient = 0.

☞ Vertical line ⇒ gradient undefined.

☞ Parallel lines have the same gradient.

Gradient formula

The gradient of the line passing through the points A(x_1, y_1) and B(x_2, y_2) is given by the formula:

$$m_{AB} = \dfrac{y_2 - y_1}{x_2 - x_1}$$

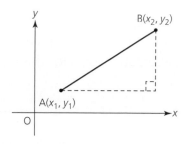

Example

Find the gradient of the line passing through the points P$(-3, 8)$ and Q$(1, 6)$.

Solution

$$m_{PQ} = \frac{6-8}{1-(-3)} = \frac{-2}{4} = \frac{-1}{2}$$

The equation of a straight line

1. $y = mx + c$

The straight line passing through the point $(0, c)$ with gradient m, has the equation: $y = mx + c$.

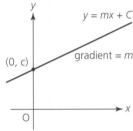

$(0, c)$ is called as the y-intercept.

Example

1 Find the equation of this straight line.

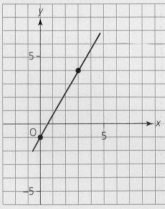

Solution

$m = \frac{5}{3}$

y-intercept $= (0, -1)$

Hence equation of line is $y = \frac{5}{3}x - 1$.

2 Find the equation of this straight line.

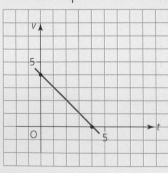

Solution

$m = \frac{-4}{4} = -1$

y-intercept $= (0, 4)$

In terms of x and y, the equation is $y = -x + 4$.
Hence in terms of v and t, the equation is $v = -t + 4$.
(*Note:* $v = 4 - t$ is an alternative form for this equation.)

2. $y - b = m(x - a)$

The straight line with gradient m passing through the point (a, b), has the equation: $y - b = m(x - a)$.

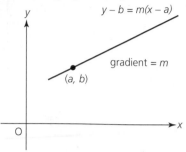

Example

1 This line has gradient $-\frac{1}{2}$. Find its equation.

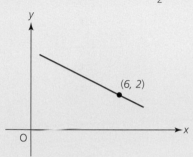

Solution

$m = -\frac{1}{2}, \quad (a, b) = (6, 2)$.

Hence equation of line is $y - 2 = -\frac{1}{2}(x - 6)$

$$\Leftrightarrow y - 2 = -\frac{1}{2}x + 3$$

$$\Leftrightarrow y = -\frac{1}{2}x + 5$$

(*Note:* $y = 5 - \frac{1}{2}x$ is an alternative form of this equation.)

2 Find the equation of this straight line.

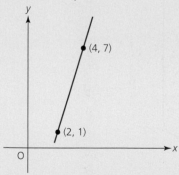

Solution

$m = \frac{7 - 1}{4 - 2} = \frac{6}{2} = 3, (a, b) = (2, 1)$.

(*Note:* $a = 4$, $b = 7$ from the point $(4, 7)$ could have been used as an alternative.)

Hence equation of line is $y - 1 = 3(x - 2)$

$$\Leftrightarrow y - 1 = 3x - 6$$

$$\Leftrightarrow y = 3x - 5$$

$\Rightarrow$

3 Find the equation of the line through $(5, -2)$ and parallel to the line with equation $y = 4x + 3$.

Solution

Parallel lines have the same gradient $\Rightarrow m = 4$.

Also $(a, b) = (5, -2)$.

Hence equation of line is $y - (-2) = 4(x - 5)$

$$\Leftrightarrow y + 2 = 4x - 20$$
$$\Leftrightarrow y = 4x - 22$$

3. Horizontal and vertical lines

The horizontal line passing through the point (a, b) has equation $y = b$.

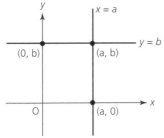

The vertical line passing through the point (a, b) has equation $x = a$.

Example

The **red** line has equation **$y = 4$**.

The **blue** line has equation **$x = -3$**.

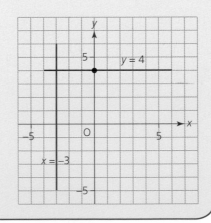

4. $ax + by + c = 0$

The equation of a straight line can also be written in the general form $ax + by + c = 0$.

This general equation can then be re-arranged so that the gradient and y-intercept can be found.

Example

Find the gradient and the *y*-intercept of the line with equation
$2x + 3y - 12 = 0$.

Solution

$2x + 3y - 12 = 0 \Leftrightarrow 3y = -2x + 12$

$$\Leftrightarrow y = -\frac{2}{3}x + 4$$

Hence the gradient $= \frac{-2}{3}$ and the *y*-intercept is $(0, 4)$.

Functional notation

Functional notation is an alternative way of expressing the relationship between two variables. For example, the equation $y = 5x - 2$ can be expressed in the form

$f(x) = 5x - 2$.

Then $f(4)$ means the value of $f(x)$ when $x = 4$.

For example: if $f(x) = 5x - 2$, then $f(4) = 5 \times 4 - 2 = 18$.

Example

A function is defined as $g(x) = 14 - 3x$.

　a) Find the value of $g(2)$.
　b) If $g(a) = -1$, find the value of *a*.

Solution

　a)　$g(x) = 14 - 3x$
　$\Rightarrow g(2) = 14 - 3 \times 2$
　　　　$= 14 - 6$
　　　　$= 8$
　b) $g(a) = -1 \Rightarrow 14 - 3a = -1$
　　　　　$\Rightarrow -3a = -1 - 14$
　　　　　$\Rightarrow -3a = -15$
　　　　　$\Rightarrow a = 5$

Linear functions

A function of the form $f(x) = ax + b$ is called a linear function since its graph is the straight line with equation $y = ax + b$.

Example

The graph of a linear function, $f(x)$, passes through the points A(4, 3) and B(8, 11).

Find a formula for $f(x)$.

Solution

$m = \frac{11-3}{8-4} = \frac{8}{4} = 2$, $(a, b) = (4, 3)$.

(*Note:* $a = 8$, $b = 11$ from the point (8, 11) could have been used as an alternative.)

Hence equation of line is $y - 3 = 2(x - 4)$

$$\Leftrightarrow y - 3 = 2x - 8$$
$$\Leftrightarrow y = 2x - 5$$

Hence a formula for the function is $f(x) = 2x - 5$.

For practice

(The answers to the following questions are given in Appendix 1.)

1 Find the gradient of the line joining the points A(7, −4) and B(19, 5).

2 Find the equation of each of these lines.

a)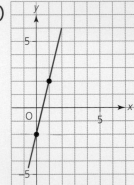

b)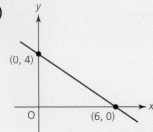

c)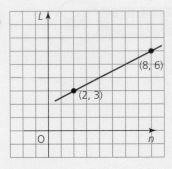

d)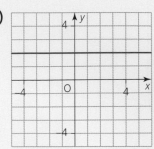

e)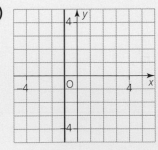

f)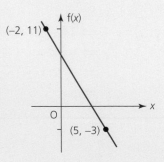

3 Find the equation of the line:
   a) through (−1, 8) and parallel to the line $y = 5x - 2$
   b) through (4, 6) and parallel to the y-axis.

4 Find the gradient and the y-intercept for each of these lines:
   a) $7x - 4y + 8 = 0$ b) $x + 2y - 10 = 0$.

5 The graph of the function $f(x) = \frac{1}{2}x + 10$ passes through the points (−6, b) and (a, 14).
   Find the values of b and a.

Chapter 7

Simultaneous equations

What you should know

You should know how to:
* ★ solve simultaneous equations graphically and algebraically
* ★ construct simultaneous equations.

To solve a pair of equations like $x + y = 11$ and $x - y = 3$, you need to find values of x and y which satisfy **both** equations simultaneously.

Pairs of equations like these are therefore called simultaneous equations.

Solving simultaneous equations graphically

To solve simultaneous equations graphically:
* draw the graph of each equation on the same diagram
  (a pair of linear equations will give two straight line graphs)
* the point of intersection of the graphs gives the solution.

Example

Solve the simultaneous equations $x + 3y = 9$ and $2x - y = 4$ graphically.

Solution

To draw the lines, first calculate the coordinates of two points on each line.

$x + 3y = 9$:   when $x = 0 \Rightarrow 3y = 9$ when $y = 0 \Rightarrow x = 9$
                            $\Rightarrow y = 3$ so $(9, 0)$ is on line
                  so $(0, 3)$ is on line

Now plot $(0, 3)$ and $(9, 0)$, then draw the line passing through these points.

$2x - y = 4$:   when $x = 0 \Rightarrow -y = 4$ when $y = 0 \Rightarrow 2x = 4$
                            $\Rightarrow y = -4$ $\Rightarrow x = 2$
                  so $(0, -4)$ is on line   so $(2, 0)$ is on the line

Now plot $(0, -4)$ and $(2, 0)$, then draw the line passing through these points.

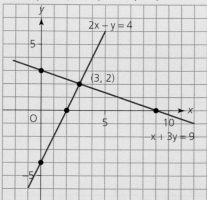

The two lines intersect at the point $(3, 2)$, so the solution of the simultaneous equations is $x = 3$, $y = 2$.

Solving simultaneous equations algebraically: substitution method

This method involves replacing y (or x) in one of the equations by an expression for y (or x) obtained from the other equation.

Example

Solve this pair of simultaneous equations: $\quad 3x + 4y = 40$

$$y = 2x - 1$$

Solution

Substitute $2x - 1$ in place of y in the first equation,

so $\qquad 3x + 4y = 40$

$\Leftrightarrow 3x + 4(2x - 1) = 40$

$\Rightarrow \quad 3x + 8x - 4 = 40$

$\Rightarrow \qquad 11x - 4 = 40$

$\Rightarrow \qquad\quad 11x = 44$

$\Rightarrow \qquad\qquad x = 4$

now substitute $x = 4$ into the second equation,

so $\quad y = 2 \times 4 - 1$

$\Rightarrow \quad y = 7$

So the solution of the simultaneous equations is $x = 4$, $y = 7$.

Solving simultaneous equations algebraically: elimination method

This is the most common method for solving simultaneous equations. It involves arranging the equations so that either the x or y coefficients are the same in both equations, and then adding or subtracting the equations to eliminate the x or the y term.

Example

Solve these pairs of simultaneous equations:

1 $\quad x + 3y = 11$
$\quad 2x - 3y = 4$

2 $\quad a + 4b = 9$
$\quad 2a + 7b = 14$

3 $7m - 3n = 17$
$\quad 5m + 4n = 6$

Solution

1 $x + 3y = 11$ (1)
$2x - 3y = 4$ (2)
add (1) + (2):
$3x = 15$
$\Rightarrow x = 5$
substitute $x = 5$ into the first equation:
$5 + 3y = 11$
$\Rightarrow 3y = 6$
$\Rightarrow y = 2$
so the solution is $x = 5, y = 2$.

2 $a + 4b = 9$ (1)
$2a + 7b = 14$ (2)
multiply (1) × 2 and leave (2) unchanged:
$2a + 8b = 18$ (3)
$2a + 7b = 14$ (2)
subtract (3) − (2):
$b = 4$
substitute $b = 4$ into equation (1):
$a + 4 \times 4 = 9$
$\Rightarrow a + 16 = 9$
$\Rightarrow a = -7$
So the solution is $a = -7, b = 4$

3 $7m - 3n = 17$ (1)
$5m + 4n = 6$ (2)
multiply (1) × 4 and (2) × 3:
$28m - 12n = 68$ (3)
$15m + 12n = 18$ (4)
add (3) + (4):
$43m = 86$
$\Rightarrow m = 2$
substitute $m = 2$ into equation (2):
$5 \times 2 + 4n = 6$
$\Rightarrow 10 + 4n = 6$
$\Rightarrow 4n = -4$
$\Rightarrow n = -1$
so the solution is $m = 2, n = -1$

Constructing and solving simultaneous equations

Many everyday problems can be solved using simultaneous equations.

Example

At a café, Rob buys 3 teas and 5 coffees for £15·75 and Morag buys 5 teas and 2 coffees for £12.

  a) Construct two equations to represent this information.

  b) Find the price of a tea and the price of a coffee.

Solution

  a) Let price of a tea = £t, and price of a coffee = £c.

$$3t + 5c = 15·75$$
$$5t + 2c = 12$$

  b)
$$3t + 5c = 15·75 \quad (1)$$
$$5t + 2c = 12 \quad (2)$$

multiply (1) × 2 and (2) × 5:

$$6t + 10c = 31·5 \quad (3)$$
$$25t + 10c = 60 \quad (4)$$

subtract (4) − (3):

$$19t = 28·5$$
$$\Rightarrow t = 1·5$$

substitute $t = 1·5$ into equation (2):

$$5 \times 1·5 + 2c = 12$$
$$\Rightarrow 7·5 + 2c = 12$$
$$\Rightarrow \quad 2c = 4·5$$
$$\Rightarrow \quad c = 2·25$$

so the price of a tea is £1·50 and the price of a coffee is £2·25.

For practice

(The answers to the following questions are given in Appendix 1.)

1 Solve the simultaneous equations $2x + y = 6$ and $x - 2y = 8$ graphically.

2 Solve this pair of simultaneous equations by substitution: $7x + 2y = 16$ $y = 5 - 2x$

3 Solve these pairs of simultaneous equations:

  a) $3x - 2y = 11$ **b)** $7x + 5y = 45$

       $7x + 8y = 13$ $4x - 3y = 14$

4 Lynn paid £26 for three adults and two children's cinema tickets.

  Scott paid £52 for four adults and seven children.

  a) Construct two equations to represent this information.

  b) Find the price of an adult's ticket and the price of a child's ticket.

5 Jim took part in a quiz. He answered 20 questions.

  He scored 5 points for each correct answer and −2 points for each incorrect answer.

  He scored a total of 51 points.

  a) Construct two equations to represent this information.

  b) Find the number of questions that Jim answered correctly and the number he answered incorrectly.

6 Find the point of intersection of the straight lines with equations

  $3x - 4y = 12$ and $2x + 5y = 31$.

Quadratic functions

What you should know

You should know how to:

★ recognise and determine the equation of a quadratic function from its graph; functions with equations expressed in the form $y = kx^2$ and $y = k(x + p)^2 + q$

★ sketch the graph of quadratic functions with equations of the form $y = (ax - m)(bx - n)$ and $y = k(x + p)^2 + q$

★ identify features of a quadratic function; nature, coordinates of turning point and equation of the axis of symmetry of quadratics of the form $y = k(x + p)^2 + q$.

A function of the form $f(x) = ax^2 + bx + c$ ($a \neq 0$) is called a quadratic function.

Its graph is a parabola with equation $y = ax^2 + bx + c$.

The graph of $f(x) = x^2$

Key points

The graph of $f(x) = x^2$ is a parabola with equation $y = x^2$.

The parabola is symmetrical about the y-axis, i.e. the equation of the axis of symmetry is $x = 0$. It has a minimum turning point at $(0, 0)$.

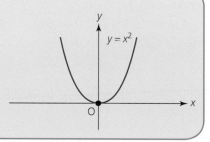

The graph of $f(x) = kx^2$

Key points

The graph of $f(x) = kx^2$ is obtained by stretching the parabola with equation $y = x^2$ parallel to the y-axis by a scale factor of k (i.e. the y coordinates of the parabola with equation $y = x^2$ are multiplied by k).

When $k < 0$, the parabola with equation $y = x^2$ is stretched by a negative scale factor parallel to the y-axis. This has the effect of reflecting the parabola with equation $y = kx^2$ in the x-axis.

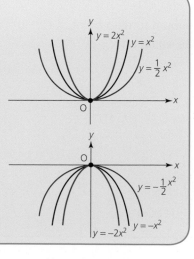

Example

A parabola of the form $y = kx^2$ passes through the point (4, 48).

Find the equation of the parabola.

Solution

Since the parabola passes through the point (4, 48), substitute

$x = 4$ and $y = 48$ into the equation $y = kx^2$:

$\Rightarrow 48 = k \times 4^2$

$\Rightarrow 48 = k \times 16$

$\Rightarrow \quad k = 3$

so the equation of the parabola is $y = 3x^2$.

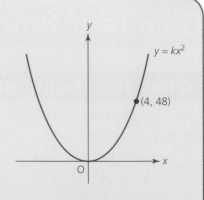

The graph of $f(x) = k\,(x + p)^2 + q$

The features of a parabola are often more easily identified when the quadratic function is expressed in completed square form
i.e. $f(x) = k\,(x + p)^2 + q$.

Key points

How to obtain the graphs of some functions related to $y = x^2$:

✳ $y = x^2 + q$ slide $y = x^2$ by q units up (parallel to the y-axis)
                      (slide down when $q < 0$)

✳ $y = (x + p)^2$ slide $y = x^2$ by p units to the left (parallel to the x-axis)
                      (slide right when $p < 0$).

When these transformations are combined and $k > 0$:

✳ $y = k\,(x + p)^2 + q$ has a minimum turning point at $(-p, q)$ and its
    axis of symmetry has equation $x = -p$.

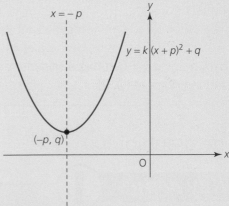

✳ $y = k\,(x - p)^2 - q$ has a minimum turning point at $(p, -q)$
    and its axis of symmetry has equation $x = p$.

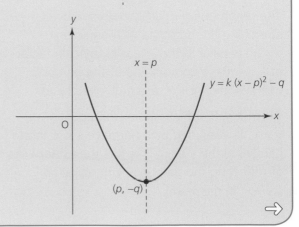

* $y = -k(x + p)^2 + q$ has a maximum turning point at $(-p, q)$ and its axis of symmetry has equation $x = -p$.

[*Note*: This equation is usually written in the form $y = q - k(x + p)^2$.]

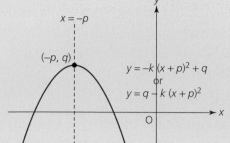

* $y = k(x - p)^2 - q$ has a maximum turning point at $(p, -q)$ and its axis of symmetry has equation $x = p$.

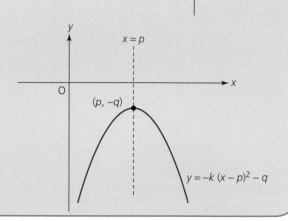

Example

1 The equations of the following parabolas are of the form $y = k(x + p)^2 + q$.

Write down the equation of each parabola and the equation of its axis of symmetry.

a)

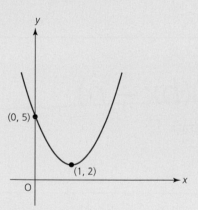

b)

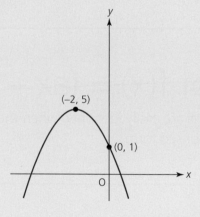

Solution

a) Turning point $(1, 2) \Rightarrow$ equation is of the form $y = k(x - 1)^2 + 2$

$(0, 5)$ on parabola $\Rightarrow 5 = k(0 - 1)^2 + 2$

$$5 = k + 2$$
$$k = 3$$

So equation of parabola is $y = 3(x - 1)^2 + 2$ and equation of axis of symmetry is $x = 1$

b) Turning point $(-2, 5) \Rightarrow$ equation is of the form $y = k(x + 2)^2 + 5$

$(0,1)$ on parabola $\Rightarrow 1 = k(0 + 2)^2 + 5$

$$1 = 4k + 5$$
$$k = -1$$

So equation of parabola is $y = -(x + 2)^2 + 5$ [usually written as $y = 5 - (x + 2)^2$]

and equation of axis of symmetry is $x = -2$.

⇨

2 Sketch the following parabolas showing the coordinates of the turning point and the y-intercept for each.

a) $y = (x + 1)^2 - 3$

b) $y = 13 - 2(x - 3)^2$

Solution

a) equation is of form $y = k(x + p)^2 + q$, $k > 0 \Rightarrow$ minimum turning point

$p = 1$, $q = -3 \Rightarrow$ turning point is $(-1, -3)$

At y-intercept, $x = 0 \Rightarrow y = (0 + 1)^2 - 3$

$$= 1 - 3$$

$$= -2 \Rightarrow \text{ the } y\text{-intercept is } (0, -2)$$

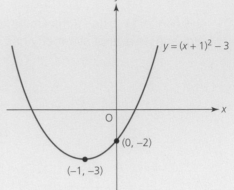

b) equation is of form $y = q - k(x + p)^2$, $k > 0 \Rightarrow$ maximum turning point

$p = -3$, $q = 13 \Rightarrow$ turning point is $(3, 13)$

At y-intercept, $x = 0 \Rightarrow y = 13 - 2(0 - 3)^2$

$$= 13 - 18$$

$$= -5 \Rightarrow \text{ the } y\text{-intercept is } (0, -5)$$

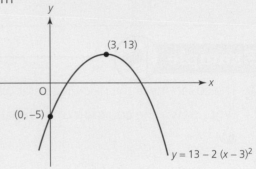

The graph of f(x) = (ax − m)(bx − n)

The x-intercepts of a parabola are easily identified when the quadratic function is expressed in factorised form.

 Key points

The graph of $y = (ax - m)(bx - n)$ crosses the x-axis at the points $(\frac{m}{a}, 0)$ and $(\frac{n}{b}, 0)$.

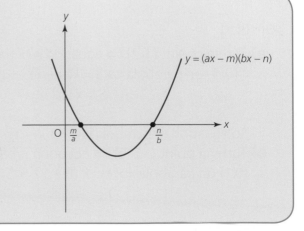

Example

1 The equation of this parabola is $y = (x + 2)(x - 4)$.

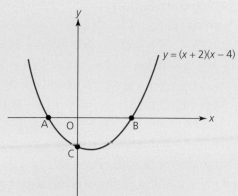

Find the coordinates of the points A, B, C and the turning point.

Solution

$y = (x - m)(x - n) = (x + 2)(x - 4) \implies m = -2, n = 4$

so A is $(-2, 0)$ and B is $(4, 0)$.

At C, $x = 0 \implies y = (0 + 2)(0 - 4)$
$\qquad\qquad\quad = 2 \times -4$
$\qquad\qquad\quad = -8$

so C is $(0, -8)$.

At the turning point, x is halfway between -2 and 4 i.e. $x = 1$

$\implies y = (1 + 2)(1 - 4)$
$\qquad = 3 \times -3$
$\qquad = -9$

so the turning point is $(1, -9)$

2 Sketch the parabola $y = (2x - 3)(2x - 7)$; showing the coordinates of the x- and y-intercepts and the turning point.

Solution

$y = (ax - m)(bx - n) = (2x - 3)(2x - 7) \implies m = 3/2, n = 7/2$

so x-intercepts are $(3/2, 0)$ and $(7/2, 0)$.

At y-intercept, $x = 0 \implies y = (2 \times 0 - 3)(2 \times 0 - 7)$
$\qquad\qquad\qquad\qquad = -3 \times -7$
$\qquad\qquad\qquad\qquad = 21$

so y-intercept is $(0, 21)$.

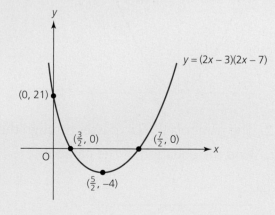

At the turning point, x is halfway between $3/2$ and $7/2$
i.e. $x = 5/2$

$\implies y = (5 - 3)(5 - 7)$
$\qquad = 2 \times -2$
$\qquad = -4$

so the minimum turning point is $(5/2, -4)$.

3 Sketch the parabola $y = (5 - x)(1 + x)$, showing the coordinates of the x- and y-intercepts and the turning point.

$y = (5 - x)(1 + x) = -(x - 5)(x + 1) \Rightarrow$ parabola has a maximum turning point

$y = -(x - m)(x - n) = -(x - 5)(x + 1) \Rightarrow m = 5, n = -1$

so x-intercepts are $(5, 0)$ and $(-1, 0)$.

At y-intercept, $x = 0 \Rightarrow y = (5 - 0)(1 + 0)$

$\qquad\qquad\qquad\qquad = 5 \times 1$

$\qquad\qquad\qquad\qquad = 5$

so y-intercept is $(0, 5)$.

At the turning point, x is halfway between -1 and 5 i.e. $x = 2$

$\Rightarrow y = (5 - 2)(1 + 2)$

$\qquad = 3 \times 3$

$\qquad = 9$

so the turning point is $(2, 9)$.

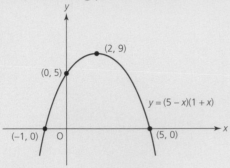

For practice

(The answers to the following questions are given in Appendix 1.)

1 The parabola $y = kx^2$ passes through the point $(3, 45)$.

Find the equation of the parabola.

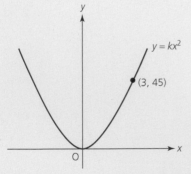

2 The parabola $y = kx^2$ passes through the point $(2, -12)$.

Find the equation of the parabola.

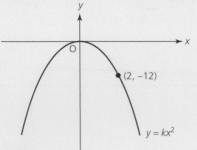

3 The equations of the following parabolas are of the form $y = k(x + p)^2 + q$.

Write down the equation of each parabola and the equation of its axis of symmetry.

a)

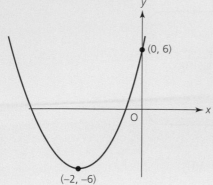

b)

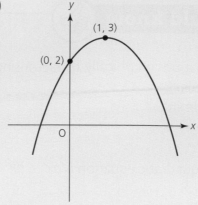

4 Sketch the following parabolas showing the coordinates of the turning point and the y-intercept for each.
   a) $y = (x - 5)^2 + 1$
   b) $y = 7 - 2(x + 4)^2$.

5 The equation of this parabola is $y = (x - 7)(x + 1)$.

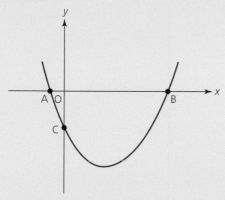

Find the coordinates of the points A, B, C and the turning point.

6 Sketch the following parabolas showing the coordinates of the x- and y-intercepts and the turning point for each.
   a) $y = (x + 3)(x - 7)$
   b) $y = (2 + x)(10 - x)$

7 Write down the equation of the axis of symmetry of the parabola $y = (4x - 1)(4x - 3)$.

Chapter 9

Quadratic equations

What you should know

You should know how to:
★ solve quadratic equations graphically, by factorising and by using the quadratic formula
★ work with the discriminant.

The standard form of a quadratic equation is $ax^2 + bx + c = 0$ ($a \neq 0$).

Solving quadratic equations graphically

Example

The graph of $y = x^2 - 2x - 3$ is shown.

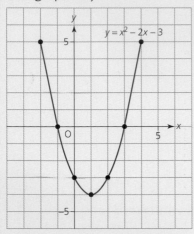

The solutions of the quadratic equation $x^2 - 2x - 3 = 0$ are $x = -1$ and $x = 3$ because these are the x values of the points where $y = 0$ on the graph.

Key points

* The solutions of the quadratic equation $ax^2 + bx + c = 0$ are the x values of the points where the graph of $y = ax^2 + bx + c = 0$ crosses the x-axis.
* The solutions are also called the roots of the equation.
* A quadratic equation can have a maximum of two roots.

Solving quadratic equations by factorising

To solve a quadratic equation by factorising:
- arrange the equation in **standard quadratic form** i.e. $ax^2 + bx + c = 0$;
  note: right side of equation *must* = 0
- factorise left side of equation;
  note: common factor, difference of two squares or trinomial
- set each factor equal to zero;
  note: product of two factors = 0 $\Rightarrow$ at least one of the factors = 0
- solve resulting equations.

Example ⚑

Solve for x:

a) $3x^2 + 15x = 0$

b) $x^2 - 16 = 0$

c) $x^2 - 5x - 24 = 0$

d) $2x^2 = 9x - 10$

Solution

a) $3x^2 + 15x = 0$

$\Rightarrow 3x(x + 5) = 0$ —— Common factor is $3x$.

$\Rightarrow 3x = 0$ or $x + 5 = 0$

$\Rightarrow x = 0$ or $x = -5$

b) $x^2 - 16 = 0$

$\Rightarrow (x + 4)(x - 4) = 0$ —— Difference of two squares.

$\Rightarrow x + 4 = 0$ or $x - 4 = 0$

$\Rightarrow x = -4$ or $x = 4$

c) $x^2 - 5x - 24 = 0$

$\Rightarrow (x + 3)(x - 8) = 0$ —— Factorise trinomial.

$\Rightarrow x + 3 = 0$ or $x - 8 = 0$

$\Rightarrow x = -3$ or $x = 8$

d) $2x^2 = 9x - 10$

$\Rightarrow 2x^2 - 9x + 10 = 0$ —— Rearrange to make right side = 0.

$\Rightarrow (2x - 5)(x - 2) = 0$ —— Factorise trinomial.

$\Rightarrow 2x - 5 = 0$ or $x - 2 = 0$

$\Rightarrow x = \frac{5}{2}$ or $x = 2$

The quadratic formula

Some quadratic expressions do not factorise. In this case the quadratic formula can be used to solve a quadratic equation.

The quadratic formula is printed in the formula list of the National 5 Mathematics paper.

The roots of the equation $ax^2 + bx + c = 0$ are given by the formula

$$x = \frac{-b \pm \sqrt{(b^2 - 4ac)}}{2a}.$$

Key points

If you are asked to give your answer to a given number of decimal places or significant figures, then you must use the formula.

Example

Find the roots of $3x^2 + 5x - 7 = 0$.

Give your answer correct to 1 decimal place.

Solution

$3x^2 + 5x - 7 = 0 \Leftrightarrow ax^2 + bx + c = 0 \Rightarrow a = 3, b = 5$ and $c = -7$

$$x = \frac{-b \pm \sqrt{(b^2 - 4ac)}}{2a} \Rightarrow x = \frac{-5 \pm \sqrt{(5^2 - 4 \times 3 \times (-7))}}{2 \times 3}$$

$$= \frac{-5 \pm \sqrt{(25 + 84)}}{6}$$

$$= \frac{-5 \pm \sqrt{109}}{6}$$

$$= \frac{-5 \pm 10 \cdot 4403}{6} \qquad \text{—— Do not round yet.}$$

$$= \frac{5 \cdot 4403}{6} \text{ or } -\frac{15 \cdot 4403}{6} \qquad \text{—— Show working.}$$

$= 0 \cdot 9067\ldots$ or $-2 \cdot 5733\ldots$

so $x = 0 \cdot 9$ or $x = -2 \cdot 6$ correct to 1 decimal place.

Quadratic equations and the discriminant

If $ax^2 + bx + c = 0 \ (a \neq 0)$, then $x = \frac{-b \pm \sqrt{(b^2 - 4ac)}}{2a}$ and $b^2 - 4ac$ is called the discriminant.

Key points

* $b^2 - 4ac > 0 \Rightarrow$ quadratic equation has two real and distinct roots.
  If the discriminant is a perfect square the roots are rational.
  If the discriminant is not a perfect square the roots are irrational.

* $b^2 - 4ac = 0 \Rightarrow$ quadratic equation has one repeated real root.
  (In this case we often say the equation has two equal real roots.)

* $b^2 - 4ac < 0 \Rightarrow$ quadratic equation has no real roots.

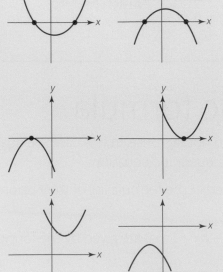

Example

1 For each equation state the nature of the roots.

   a) $x^2 - 2x - 5 = 0$

   b) $2x^2 + x + 3 = 0$.

Solution

   a) $a = 1, b = -2, c = -5$

$$\Rightarrow b^2 - 4ac = (-2)^2 - 4 \times 1 \times (-5)$$
$$= 4 + 20$$
$$= 24$$

$$\Rightarrow b^2 - 4ac > 0 \text{ and not a perfect square}$$
$$\Rightarrow \text{two real and distinct, irrational roots.}$$

   b) $a = 2, b = 1, c = 3$

$$\Rightarrow b^2 - 4ac = 1^2 - 4 \times 2 \times 3$$
$$= 1 - 24$$
$$= -23$$

$$\Rightarrow b^2 - 4ac < 0$$
$$\Rightarrow \text{no real roots}$$

2 State the nature of the roots of $(5x - 1)^2 = 10x - 3$.

Solution

$$(5x - 1)^2 = 10x - 3 \Rightarrow 25x^2 - 10x + 1 = 10x - 3$$
$$\Rightarrow 25x^2 - 20x + 4 = 0$$

$$a = 25, b = -20, c = 4$$
$$\Rightarrow b^2 - 4ac = 20^2 - 4 \times 25 \times 4$$
$$= 400 - 400$$
$$= 0$$
$$\Rightarrow \text{two equal real roots}$$

Using quadratic equations to solve problems

Example

1 A rectangular lawn is $(x + 2)$ metres long and $(x - 1)$ metres wide.
The area of the lawn is 40 square metres.

   a) Show that $x^2 + x - 42 = 0$.

   b) Find the length of the lawn.

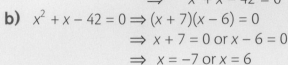

$(x + 2)$ m

$(x - 1)$ m

Solution

   a) $A = (x + 2)(x - 1) \Rightarrow (x + 2)(x - 1) = 40$

   $\Rightarrow x^2 - x + 2x - 2 = 40$

   $\Rightarrow x^2 + x - 2 = 40$

   $\Rightarrow x^2 + x - 42 = 0$

   b) $x^2 + x - 42 = 0 \Rightarrow (x + 7)(x - 6) = 0$

   $\Rightarrow x + 7 = 0$ or $x - 6 = 0$

   $\Rightarrow x = -7$ or $x = 6$

   but $x \neq -7$ since it is impossible to have a lawn with a negative length

   so $x = 6$

   hence the length of the lawn $= x + 2$

   $= 6 + 2$

   $= 8$ metres

2 Find the coordinates of the points of intersection of the parabola $y = x^2 + 5x - 7$ and the line $y = 2x + 3$.

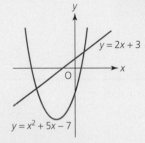

Solution

At the points of intersection $\quad x^2 + 5x - 7 = 2x + 3$

$\Rightarrow \quad x^2 + 3x - 10 = 0$

$\Rightarrow \quad (x + 5)(x - 2) = 0$

$\Rightarrow x + 5 = 0$ or $x - 2 = 0$

$\Rightarrow \quad x = -5$ or $x = 2$

To find the corresponding y-coordinates of the points of intersection, substitute $x = -5$ and $x = 2$ into the equation $y = 2x + 3$.

when $x = -5, y = 2 \times -5 + 3$ $\qquad$ when $x = 2, y = 2 \times 2 + 3$

$\Rightarrow \qquad y = -7$ $\qquad\qquad \Rightarrow \qquad y = 7$

so the points of intersection are $(-5, -7)$ and $(2, 7)$.

For practice

(The answers to the following questions are given in Appendix 1.)

1 Use the graphs below to find the roots of
   a) $x^2 + x - 6 = 0$
   b) $4 + 3x - x^2 = 0$

a)

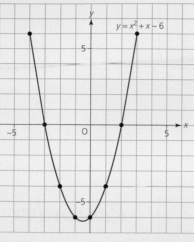

b)

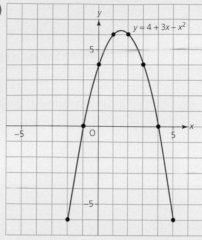

2 Solve for x:
   a) $x^2 - 8x = 0$
   b) $4x^2 - 9 = 0$
   c) $x^2 - 12x + 36 = 0$
   d) $x^2 = 3x + 28$
   e) $3x^2 + 5x + 2 = 0$

3 Find the roots of the following equations.

Give your answer correct to 1 decimal place.
   a) $4x^2 + 3x - 2 = 0$
   b) $x^2 - 5x + 3 = 0$

4 For each of the following equations state the nature of the roots.
   a) $9x^2 - 12x - 24 = 0$
   b) $3x^2 + 2x + 1 = 0$

5 State the nature of the roots of the equation $16x^2 + 8x + 1 = 0$.

6 State the nature of the roots of the equation $(2x + 1)(x - 3) = 0$.

7 A triangle with base $2x$ cm and height $(x + 4)$ cm has an area of 45 cm^2.

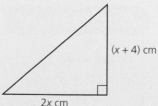

   a) Show that $x^2 + 4x - 45 = 0$.
   b) Find the length of the base of the triangle.

8 Find the coordinates of the points of intersection of the parabola $y = x^2 + x + 2$ and the line $y = 3x + 5$.

Chapter 10
Pythagoras' Theorem

What you should know

You should know how to:
* ★ use Pythagoras' Theorem in complex situations (including 3-D)
* ★ use the converse of Pythagoras' Theorem.

When we know the lengths of two sides of a right-angled triangle, we can use Pythagoras' Theorem to find the length of the third side.

Key points

Pythagoras' Theorem states that:

If triangle ABC is right-angled at C, then $c^2 = a^2 + b^2$.

To find the length of one of the two shorter sides then rearrange to give:

$a^2 = c^2 - b^2$
or $b^2 = c^2 - a^2$

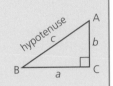

Remember

The **hypotenuse** is the longest side of a right-angled triangle; it is always opposite the right angle.

Example

Calculate the height of this isosceles triangle ABC.

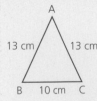

Solution

Draw line AD which represents the height of the triangle.
Use Pythagoras' Theorem in right-angled triangle ABD.

AD is one of the shorter sides of the right-angled triangle, so use $b^2 = c^2 - a^2$.

$AD^2 = 13^2 - 5^2$
$= 169 - 25$
$= 144$
$\Rightarrow AD = \sqrt{144}$
$= 12$

so the height of the triangle is 12 cm.

Hints & tips

Do **not** write $x^2 = 13^2 + 5^2$.

x is not the length of the **hypotenuse**, so start with $x^2 = 13^2 - 5^2$.

Example

1 Calculate the length of the line joining the points P(−3, −2) and Q(5, 2).
Give the answer correct to 1 decimal place.

Solution

Plot P and Q as shown.

Draw in lines PQ, PR and QR to form right-angled triangle PQR.
PQ is the hypotenuse of the right-angled triangle, so use $c^2 = a^2 + b^2$.

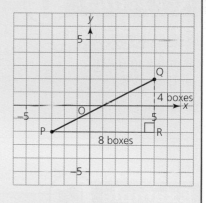

$$PQ^2 = PR^2 + QR^2$$
$$= 8^2 + 4^2$$
$$= 64 + 16$$
$$= 80$$
$$\Rightarrow PQ = \sqrt{80}$$
$$= 8.944$$
$$= 8.9 \text{ to 1 decimal place.}$$

2 ABCDEFGH is a cuboid as shown.
Calculate the length of space diagonal AG.
Give the answer correct to 2 significant figures.

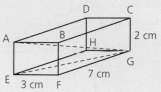

Solution

Triangle EFG is right-angled at F.

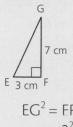

$$EG^2 = FF^2 + FG^2$$
$$= 3^2 + 7^2$$
$$= 9 + 49$$
$$= 58$$
$$\Rightarrow EG = \sqrt{58}$$ —— For accuracy the exact value $\sqrt{58}$ should be used at this stage.

Triangle AEG is right-angled at E.

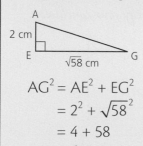

$$AG^2 = AE^2 + EG^2$$
$$= 2^2 + \sqrt{58}^2$$
$$= 4 + 58$$
$$= 62$$
$$\Rightarrow AG = \sqrt{62}$$
$$= 7.874....$$
$$= 7.9 \text{ to 2 significant figures.}$$

The converse of Pythagoras' Theorem

When we know the lengths of all three sides of a triangle, we can use the converse of Pythagoras' Theorem to determine if the triangle is right-angled.

Key points

The converse of Pythagoras' Theorem states that:

If in triangle ABC, $c^2 = a^2 + b^2$, then the triangle is right-angled at C.

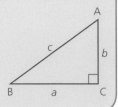

Example

1 Prove that triangle DEF is right-angled.

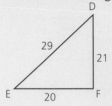

Solution

longest side: $DE^2 = 29^2$
$= 841$

shorter sides: $EF^2 + FD^2 = 20^2 + 21^2$
$= 400 + 441$
$= 841$
$\Rightarrow DE^2 = EF^2 + FD^2$

so triangle DEF is right-angled, by the converse of Pythagoras' Theorem.

2 Is triangle STU right-angled?

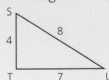

Solution

longest side: $SU^2 = 8^2$
$= 64$

shorter sides: $ST^2 + TU^2 = 4^2 + 7^2$
$= 16 + 49$
$= 65$
$\Rightarrow SU^2 \neq ST^2 + TU^2$

so triangle STU is not right-angled.

Hints & tips

1 Do not start by writing $10^2 + 8^2 = 13^2$ as this may not be true.
In fact, $10^2 + 8^2 = 164$ and $13^2 = 169$, so $10^2 + 8^2 \neq 13^2$.

2 Never state that $AB^2 = AC^2 + BC^2$ or $AB^2 \neq AC^2 + BC^2$ until you have carried out the calculations necessary to justify your conclusion.

For practice

(The answers to the following questions are given in Appendix 1.)

1 Calculate the value of *x* in each diagram.
Round the answer to 1 decimal place where necessary.

a)

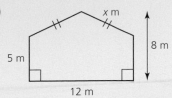

b)

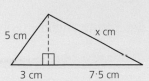

2 Calculate the distance between the points U(−4, 1) and V(3, −5).
Give the answer correct to 2 significant figures.

3 A cone has perpendicular height 20 cm and slant height 25 cm.

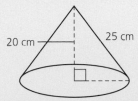

Calculate the diameter of the cone.

4 CDEFGHIJ is a cube of edge 8cm.

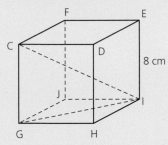

Calculate the length of space diagonal CI.
Give the answer correct to 3 significant figures.

5 Prove these triangles are right-angled or not.

a)

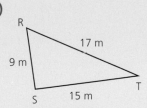

b)

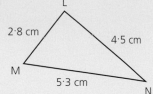

6 Prove that ABCD is a rectangle.

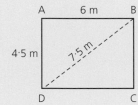

A polygon is a two dimensional shape with straight sides.

A three sided polygon is a triangle.

A four sided polygon is a quadrilateral.

In a regular polygon all the sides are equal and all the angles are equal.

Triangles

The sum of the angles of a triangle is 180°.

An **equilateral triangle** has three equal sides and three equal angles,
each 60°.

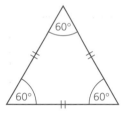

An **isosceles triangle** has two equal sides and two equal angles.

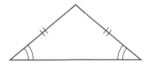

Quadrilaterals

Some examples of quadrilaterals are: square, rectangle, rhombus, kite,
parallelogram and trapezium. The sum of the angles of a quadrilateral is 360°.

A **rhombus** has:
● four equal sides
● opposite sides which are parallel
● opposite angles which are equal.

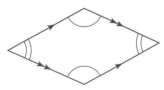

A **kite** has:
● two pairs of adjacent sides which are equal
● one pair of opposite angles which are equal.

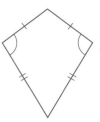

A **parallelogram** has:
- opposite sides which are equal and parallel
- opposite angles which are equal.

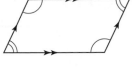

A **trapezium** has:
- one pair of opposite sides which are parallel.

Corresponding, alternate and allied (co-interior) angles

Corresponding angles

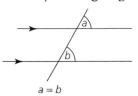

$a = b$

Alternate angles

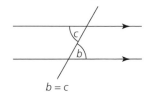

$b = c$

Allied (or co-interior) angles

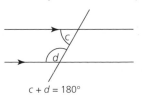

$c + d = 180°$

Example

1 In the diagram:
- AB is parallel to CE
- BDE is an isosceles triangle
- angle DAB = 65° and angle DBE = 70°.

Calculate the size of angle ADB.

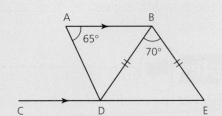

Solution

∠CDA = 65° ———— Alternate to ∠DAB.

∠BDE = 55° ———— Isosceles triangle ⇒ ∠BDE = 1/2 of (180 − ∠DBE).

⇒ ∠ADB = 60° ———— Straight angle ⇒ ∠ADB = 180 − ∠CDA − ∠BDE.

2 In the diagram:
- PU is parallel to QT
- QRST is a kite
- angle PUT = 80° and angle TQR = 125°.

Calculate the size of angle QRS.

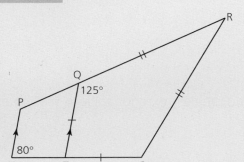

Solution

∠QTS = 80° ———— Corresponding to ∠PUT.

∠TSR = 125° ———— Kite ⇒ ∠TSR = ∠TQR.

⇒ ∠QRS = 30° ———— Kite ⇒ ∠QRS = 360 − ∠QTS − ∠TSR − ∠TQR.

Angles in polygons

Angles formed by the sides of a polygon are called **interior angles**. When a side of a polygon is extended, as shown, the angle formed is called an **exterior angle**.

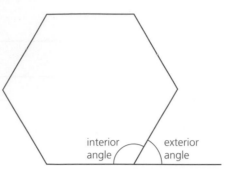

interior angle
exterior angle

At each vertex of the polygon:
interior angle + exterior angle = 180°.

When the diagonals from one vertex of an n-sided polygon are drawn, $(n - 2)$ triangles are formed.

a)

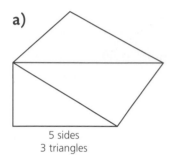

5 sides
3 triangles

b)

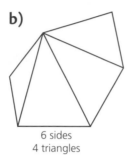

6 sides
4 triangles

So, the sum of the interior angles is given by $(n - 2) \times 180°$.

The angles of a regular polygon are all equal, so the interior angle of a regular n-sided polygon is given by $\dfrac{(n - 2) \times 180°}{n}$.

> ## Example 🚩
>
> Find the sizes of the interior and exterior angles of a regular pentagon.
>
> ### Solution
> a pentagon has 5 sides $\Rightarrow$ interior angle $= \dfrac{180° \times (5 - 2)}{5}$
>
> $$= \dfrac{180° \times 3}{5}$$
>
> $$= \dfrac{540°}{5}$$
>
> $$= 108°$$
>
> interior angle + exterior angle = 180° $\Rightarrow$ exterior angle
> $$= 180° - \text{interior angle}$$
> $$= 180° - 108°$$
> $$= 72°$$

For practice

(The answers to the following questions are given in Appendix 1.)

1 In the diagram:
   - ABCD is a square
   - DB is a diagonal of the square
   - AFD is an equilateral triangle.

   Calculate the size of angle BEF.

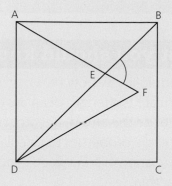

2 In the diagram:
   - PQRS is a parallelogram
   - angle TSR = 105°, angle SUR = 115° and angle UPQ = 40°.

   Calculate the size of angle PSU.

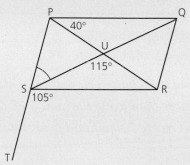

3 In the diagram:
   - EDFG is a kite
   - EDFH is a rhombus
   - angle EGF = 50° and angle GEH = 35°.

   Calculate the size of angle EDF.

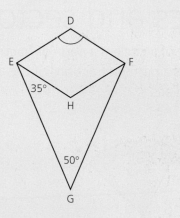

4 In the diagram:
   - YZ is parallel to WV
   - YX is parallel to ZW
   - angle ZYX = 50° and angle YXW = 60°.

   Calculate the size of angle VWX.

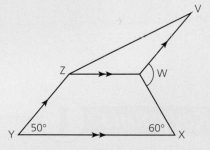

5 Find the sizes of the interior and exterior angles of a regular octagon.

Circles

Remember

The formula for the circumference of a circle is $C = \pi d$ and for the area of a circle is $A = \pi r^2$.

Hints & tips ⭐

Using the π button on your calculator will always give you a more accurate answer than using the value 3·14. Give the answers to calculations involving π correct to 3 significant figures unless instructed otherwise.

Arcs and sectors

An arc of a circle is a part (or fraction) of the circumference. A sector of a circle is a part (or fraction) of the area of the circle, bounded by two radii and an arc.

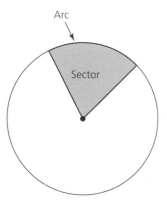

Key points ⚠

For a sector with angle $x°$ at the centre:

* length of arc $= \frac{x}{360} \times \pi d$

* area of sector $= \frac{x}{360} \times \pi r^2$

To find the angle at the centre:

* $x° = \frac{\text{length of arc}}{\pi d} \times 360°$ ——— If you know the arc length and radius.

* $x° = \frac{\text{area of sector}}{\pi r^2} \times 360°$ ——— If you know the sector area and radius.

Example

Find:

a) the length of arc AB

b) the area of sector AOB.

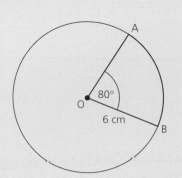

Solution

a) length of arc $= \frac{80}{360} \times \pi \times 12$

$= 8.377...$

$= 8.38\text{cm}$ (to 3 s.f.)

b) area of sector $= \frac{80}{360} \times \pi \times 6^2$

$= 25.132...$

$= 25.1\text{cm}^2$ (to 3 s.f.)

Example

a)

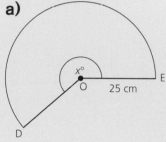

b)

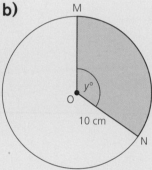

Find the size of angle:

a) x, given that arc DE = 96 cm and the radius of the circle is 25 cm.

b) y, given that area of shaded sector = 109 cm^2 and the radius of the circle is 10 cm.

Give the answers correct to 3 significant figures.

Solution

a) $x° = \frac{\text{length of arc}}{\pi d} \times 360°$

$= \frac{96}{\pi \times 50} \times 360°$

$= \frac{96}{157 \cdot 079} \times 360°$

$= 220 \cdot 015...$

$= 220°$ (to 3 s.f.)

b) $y° = \frac{\text{area of sector}}{\pi r^2} \times 360°$

$= \frac{109}{\pi \times 10^2} \times 360°$

$= \frac{109}{314 \cdot 159} \times 360°$

$= 124 \cdot 904...$

$= 125°$ (to 3 s.f.)

The circle questions that you will meet in the National 5 examination will often test your reasoning (problem solving) skills.

Example

The shape in the diagram consists of a sector of a circle and a kite OABC.

- The circle has radius 7 cm.
- Angle AOC = 130°.
- AB = AC = 15 cm.

Find the perimeter of the shape.

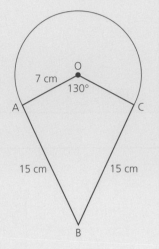

Solution

Angle at centre of sector = 360° − 130°

$$= 230°$$

$$\Rightarrow \text{length of arc} = \tfrac{230}{360} \times \pi \times 14$$

$$= 28{\cdot}099\ldots$$

$$\Rightarrow \text{perimeter of shape} = 28{\cdot}099 + 15 + 15$$

$$= 58{\cdot}1 \text{ cm (to 3 s.f.)}$$

Example

The diagram shows a quadrant of a circle.

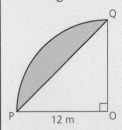

Find the area of the shaded segment.

Solution

area of segment = area of sector POQ − area of triangle POQ.

area of sector POQ $= \tfrac{90}{360} \times \pi \times 12^2$ area of triangle POQ $= \tfrac{1}{2} \times 12 \times 12$

$= 113{\cdot}097\ldots$ $= 72$

$$\Rightarrow \text{area of segment} = 113{\cdot}097\ldots - 72$$

$$= 41{\cdot}097\ldots$$

$$= 41{\cdot}1 \text{ m}^2 \text{ (to 3 s.f.)}$$

Circle geometry

There are some properties of circles that you are assumed to know already.

Remember

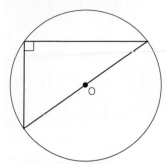

The angle in a semi-circle is a right angle.

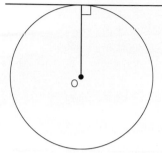

A tangent is perpendicular to the radius at the point of contact.

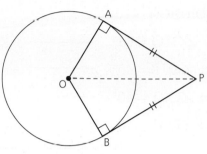

Tangents drawn to a circle from the same point are equal in length. Quadrilateral OAPB is called a tangent kite.

Example

In the diagram:
- BC is a diameter of the circle with centre O
- A is a point on the circumference of the circle.

Find the area of the circle.

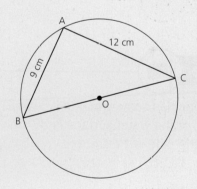

Solution

$\angle BAC = 90°$ ⎯⎯⎯⎯ Angle in a semi-circle.

$\Rightarrow BC^2 = BA^2 + AC^2$ ⎯⎯⎯ Pythagoras' Theorem.

$\quad\quad = 9^2 + 12^2$

$\quad\quad = 81 + 144$

$\quad\quad = 225$

$\Rightarrow BC = \sqrt{225}$

$\quad\quad = 15$

$\Rightarrow$ radius $= 15 \div 2 = 7 \cdot 5$

$\quad \Rightarrow$ area $= \pi r^2$

$\quad\quad\quad = \pi \times 7 \cdot 5^2$

$\quad\quad\quad = 176 \cdot 71$

$\quad\quad\quad = 177 \text{ cm}^2$ (to 3 s.f.)

Example

In the diagram:

- O is the centre of the circle
- PQ is a tangent to the circle at T
- angle RTQ = 50°.

Calculate the size of angle ROT.

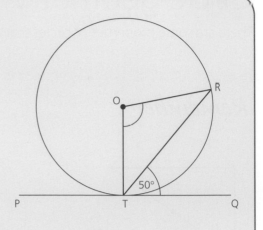

Solution

$\angle OTQ = 90°$ ———— Point of contact of tangent and radius.

$\Rightarrow \angle OTR = 40°$ ———— $\angle OTQ - 50°$.

$\Rightarrow \angle ORT = 40°$ ———— Isosceles triangle $\Rightarrow \angle ORT = \angle OTR$.

$\Rightarrow \angle ROT = 100°$ ———— Sum of angles in triangle ROT = 180°.

Perpendicular bisector of a chord

A chord is a line joining two points on the circumference of a circle.

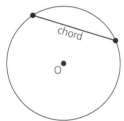

Key points

The perpendicular bisector of a chord passes through the centre of the circle.

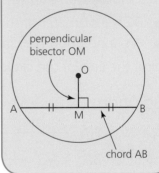

perpendicular
bisector OM

chord AB

Example

A cylindrical pipe has some water in it.
The width, AB, of the water surface is 30 cm.
The radius of the pipe is 17 cm.
Find the depth, MC, of the water.

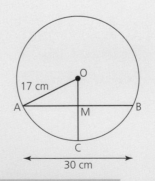

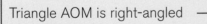

Solution

Triangle AOM is right-angled ———— Perpendicular bisector of chord.

$\Rightarrow OM^2 = OA^2 - AM^2$ ———— Pythagoras' Theorem.

$\qquad = 17^2 - 15^2$

$\qquad = 289 - 225$

$\qquad = 64$

$\Rightarrow OM = \sqrt{64}$

$\qquad = 8$

$\Rightarrow$ depth of water, MC = OC − OM

$\qquad\qquad = 17 - 8$ ———— OC = radius.

$\qquad\qquad = 9$ cm

For practice

(The answers to the following questions are given in Appendix 1.)

1 Find the length of minor arc AB.

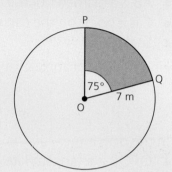

2 Find the area of shaded sector POQ.

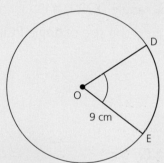

3 In the diagram, the radius of the circle is 9 cm and the length of arc DE is 11 cm.

Find the size of angle DOE.

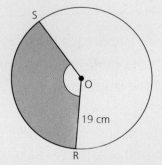

4 In the diagram, the radius of the circle is 19 cm and the area of shaded sector SOR is 441 cm^2.

Find the size of angle SOR.

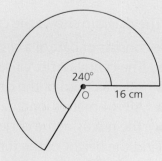

5 The shape shown is a sector of a circle, centre O and radius 16 cm.
Find the perimeter of the shape.

6 RS and TU are arcs of a circle, centre O.

OR = 8 m, OU = 5 m and angle TOU = 50°.

Find the shaded area.

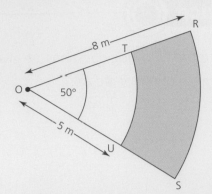

7 In the diagram:
- AB is a diameter of the circle with centre O
- C is a point on the circumference of the circle
- angle CAO = 37°.

Find the size of angle OCB.

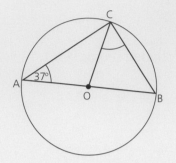

8 In the diagram:
- OPQR is a tangent kite
- angle PQR = 44°.

Find the size of angle POR.

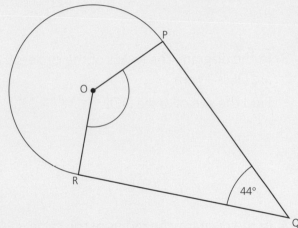

9 Two circles each have centre O.

XY is a tangent to the smaller circle and a chord of the larger circle.

XY is 20 cm.

The radius of the smaller circle is 7·5 cm.

Find the radius of the larger circle.

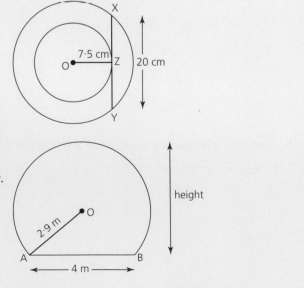

10 The diagram shows the cross-section of a tunnel.

It consists of part of a circle with a horizontal base.
- The centre of the circle is O.
- AB is a chord of the circle.
- AB is 4 metres long.
- The radius of the circle is 2·9 metres.

Find the height of the tunnel.

What you should know

You should know how to:
★ work with similar figures; interrelationship of length, area and volume.

Two figures are **similar** when they are the same shape, with one being an enlargement or a reduction of the other.

Key points

Similar shapes have their:
* corresponding angles equal ———— The figures are said to be **equiangular** when this is true.
* corresponding sides in the same ratio. ———— The ratio is known as the **scale factor**.

The scale factor is:
* greater than 1 for an enlargement
* less than 1 (but greater than 0) for a reduction.

Example

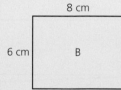

Rectangles A and B are similar.

B is an enlargement of A – the scale factor is 2. ———— Note: 6:3 = 8:4 = 2:1.

A is a reduction of B – the scale factor is $\frac{1}{2}$. ———— Note: 3:6 = 4:8 = 1:2.

Example

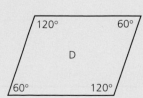

Parallelograms C and D are not similar since their corresponding angles are not equal (i.e. they are not equiangular).

Similar triangles

Example

In the diagram DE is parallel to BC.

a) Explain why triangle ABC is similar to triangle ADE.

b) Calculate the length of AB.

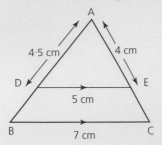

Solution

a) ∠ABC = ∠ADE ——————— Corresponding angles

∠ACB = ∠AED ——————— Corresponding angles

∠BAC = ∠DAE ——————— Same angle

⇒ triangles ABC and ADE are equiangular

⇒ triangles ABC and ADE are similar.

b) scale factor = $\frac{BC}{DE}$ = $\frac{7}{5}$ ——— Enlargement ⇒ $\frac{\text{large length}}{\text{small length}}$

⇒ AB = $\frac{7}{5}$ × 4.5 = 6.3 cm

Key points

When two triangles are equiangular their corresponding sides are in the same ratio and the triangles are similar.

Example

In the diagram PQ is parallel to ST.

a) Explain why triangle PQR is similar to triangle RST.

b) Calculate the length of RS.

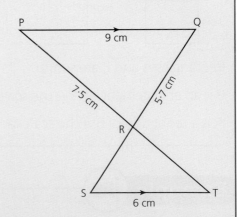

Solution

a) ∠QPR = ∠RTS ——— Alternate angles

∠PQR = ∠RST ——— Alternate angles

∠PRQ = ∠SRT ——— Vertically opposite angles

⇒ triangles PQR and RST are equiangular

⇒ triangles PQR and RST are similar.

b) scale factor = $\frac{ST}{PQ}$ = $\frac{6}{9}$ — Reduction ⇒ $\frac{\text{small length}}{\text{large length}}$

⇒ RS = $\frac{6}{9}$ × 5.7 = 3.8 cm

Areas of similar figures

Example 🚩

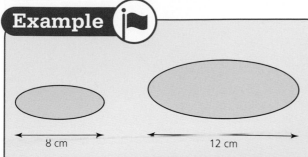

8 cm   12 cm

These two ellipses are mathematically similar.

The width of the smaller one is 8 cm and the width of the larger one is 12 cm.

The area of the smaller ellipse is 72 cm².

Find the area of the larger ellipse.

Solution

Scale factor $= \frac{12}{8} = \frac{3}{2}$

$\Rightarrow$ area of larger ellipse $= \left(\frac{3}{2}\right)^2 \times 72 = 162$ cm²

Key points ❗

If two figures are similar and the length scale factor is k, then the area scale factor is k^2.

Volumes of similar figures

Example 🚩

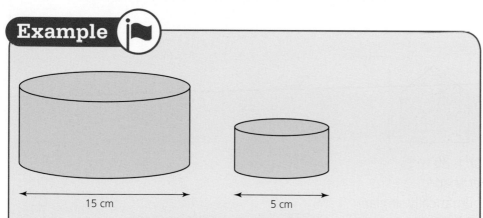

15 cm   5 cm

These two cylinders are mathematically similar.

The diameter of the larger one is 15 cm and the diameter of the smaller one is 5 cm.

The volume of the larger cylinder is 2160 cm³.

Find the volume of the smaller cylinder.

Solution

Scale factor $= \frac{5}{15} = \frac{1}{3}$

$\Rightarrow$ volume of smaller cylinder $= \left(\frac{1}{3}\right)^3 \times 2160 = 80$ cm³

Key points ❗

If two objects are similar and the length scale factor is k, then the volume scale factor is k^3.

For practice

(The answers to the following questions are given in Appendix 1.)

1 In the diagram VZ is parallel to WY.
Calculate the length of WX.

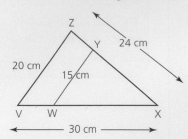

2 In the diagram DE is parallel to GH.
Calculate the length of DF.

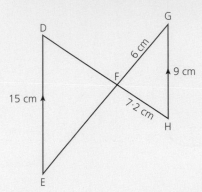

3 These two shapes are mathematically similar.
The height of the larger shape is 5 m and the height of the smaller one is 3 m.

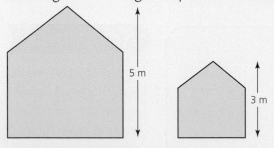

The area of the larger shape is 30 m².
Find the area of the smaller shape.

4 These two cones are mathematically similar.
The diameter of the smaller one is 9 cm and the diameter of the larger one is 18 cm.

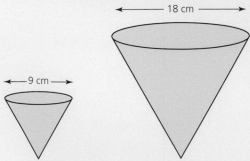

The volume of the smaller cone is 150 cm³.
Find the volume of the larger cone.

5 Shown are two slices of pizza which are mathematically similar.

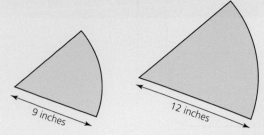

9 inches

12 inches

The price of the pizza depends only on the area of each slice.
Find the price of the large slice if the price of the small slice is £1·35.

6 Shown are two cartons of ice cream which are mathematically similar.

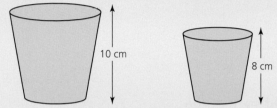

10 cm

8 cm

The price of the ice cream depends only on the volume of ice cream in the carton.
Find the price of the small carton of ice cream if the price of the large carton is £1·95.
Give the answer correct to the nearest penny.

Volume is the amount of space occupied by a three-dimensional shape.

Volume is usually measured in cubic centimetres (cm^3) or cubic metres (m^3).

Remember

- Volume of a cuboid = length × breadth × height ⇔ $V = lbh$

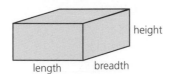

- Volume of a cube = length × length × length ⇔ $V = l^3$

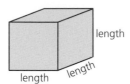

- Volume of a prism = area of cross-section × height
  or area of cross-section × length
  ⇔ $V = Ah$ or $V = Al$

  Hexagonal prism   Triangular prism

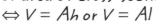
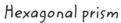

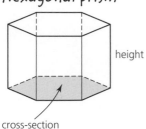

cross-section   cross-section

- A cylinder is a prism whose cross-section is a circle
  ⇒ volume of a cylinder is given by $V = \pi r^2 h$

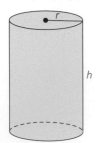

Example

Calculate the volume of this triangular prism.

Give your answer in m^3.

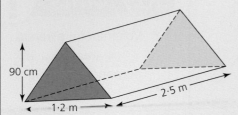

Solution

Area of cross-section $= \frac{1}{2}bh = \frac{1}{2} \times 1.2 \times 0.9 = 0.54$ m^2

Volume of prism $= Al = 0.54 \times 2.5 = 1.35$ m^3.

Example

Calculate the volume of this cylinder.

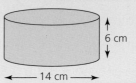

Solution

$V = \pi r^2 h = \pi \times 7^2 \times 6$

$\quad = 923.62\ldots$

$\quad = 924$ cm^3 (to 3 s.f.)

Example

A cylindrical jar has a diameter of 10 cm.
1·2 litres of water is poured into the jar.
Find the depth of the water in the jar.

Solution

$V = 1200$ ———————— 1·2 litres = 1200 cm^3

$V = \pi r^2 h$

$\Rightarrow 1200 = \pi \times 5^2 \times d$ —— d = depth of water in jar

$1200 = 78.53\ldots \times d$

$\Rightarrow d = \dfrac{1200}{78.53\ldots}$

$\quad = 15.27$

$\quad = 15.3$ cm (to 3 s.f.)

Remember

1 litre = 1000 ml = 1000 cm^3.

Example

The end of this prism consists of a rectangle and a quarter circle.
Find the volume of the prism.

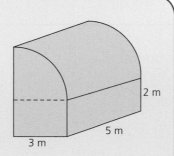

Solution

Area of cross-section $= lb + \frac{1}{4}\pi r^2$

$\qquad\qquad\qquad = 3 \times 2 + \frac{1}{4} \times \pi \times 3^2$

$\qquad\qquad\qquad = 6 + 7.068\ldots$

$\qquad\qquad\qquad = 13.068\ldots$ m^2

Volume of prism $= Al = 13.068 \times 5$

$\qquad\qquad\qquad = 65.342\ldots$

$\qquad\qquad\qquad = 65.3$ m^3 (to 3 s.f.)

Volume of pyramid, cone and sphere

Find the volume of this square-based pyramid.

Solution

Area of base = $6 \times 6 = 36$ cm^2

$\Rightarrow V = \frac{1}{3}Ah = \frac{1}{3} \times 36 \times 8 = 96$ cm^3.

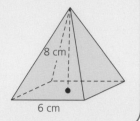

A sphere of diameter 8 cm has the same volume as a cone with height 12 cm.

Find the diameter of the cone.

Solution

Volume of sphere = $\frac{4}{3}\pi r^3 = \frac{4}{3} \times \pi \times 4^3$

$= \frac{4}{3} \times \pi \times 64$

$= 268 \cdot 08\ldots$ cm^3

Volume of cone = $\frac{1}{3}\pi r^2 h \Rightarrow 268 \cdot 08\ldots = \frac{1}{3} \times \pi \times r^2 \times 12$

$268 \cdot 08\ldots = 12 \cdot 56\ldots \times r^2$

$\Rightarrow \qquad r^2 = \dfrac{268 \cdot 082\ldots}{12 \cdot 566\ldots}$

$\Rightarrow \qquad r = \sqrt{21 \cdot 333}$

$\qquad = 4 \cdot 618\ldots$

$\Rightarrow$ diameter = $2 \times 4 \cdot 618\ldots$

$\qquad = 9 \cdot 237$

$\qquad = 9 \cdot 24$ cm (to 3 s.f.)

* Volume of a pyramid
  $= \frac{1}{3} \times$ area of base $\times$ perpendicular height
  $\Leftrightarrow V = \frac{1}{3}Ah$

* A cone is a pyramid whose base is a circle
  $\Leftrightarrow$ Volume of a cone is given by $V = \frac{1}{3}\pi r^2 h$

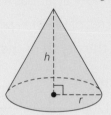

* Volume of a sphere is given by $V = \frac{4}{3}\pi r^3$

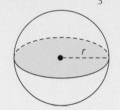

This object consists of a cone on top of a hemisphere.

Calculate the volume of the object.

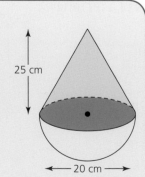

Solution

Volume of hemisphere = $\frac{2}{3}\pi r^3 = \frac{2}{3} \times \pi \times 10^3$

$\qquad = 2094 \cdot 39\ldots$

Volume of cone = $\frac{1}{3}\pi r^2 h = \frac{1}{3} \times \pi \times 10^2 \times 25$

$\qquad = 2617 \cdot 99\ldots$

$\Rightarrow$ Volume of object = $2094 \cdot 39\ldots + 2617 \cdot 99\ldots = 4712 \cdot 38\ldots$

$\qquad = 4710$ cm^3 (to 3 s.f.)

For practice

(The answers to the following questions are given in Appendix 1.)

1 The end section of a building consists of a rectangle and a semi-circle.

Calculate the volume of the building.

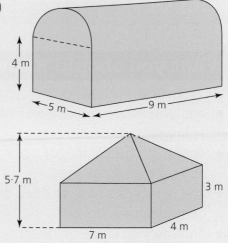

4 m

5 m

9 m

2 This building consists of a cuboid with a pyramid on top.
Calculate the volume of the building.

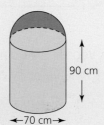

5·7 m

3 m

7 m

4 m

3 This water tank consists of a hemisphere on top of a cylinder.
How many litres of water will the tank hold when full?

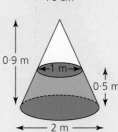

90 cm

70 cm

4 This podium is formed by slicing a small cone off the top of a large cone.
Calculate the volume of the podium.

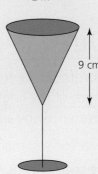

0·9 m

1 m

0·5 m

2 m

5 This conical glass can hold 150 millilitres of liquid when full.
Calculate the diameter of the top of the glass.

9 cm

6 A metal cube of side 5 cm is melted down and made into spheres of radius 1 cm.
How many complete spheres can be made from the melted down cube?

What you should know

You should know how to:
- ★ add or subtract two-dimensional vectors using directed line segments
- ★ determine coordinates of a point from a diagram representing a three-dimensional object
- ★ add or subtract two- or three-dimensional vectors using components
- ★ find the magnitude of a vector.

A vector is a quantity which has both magnitude (length) and direction, e.g. the velocity of a wind blowing from the north east at 30 mph.

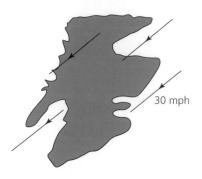

30 mph

Directed line segments

Vectors can be represented in diagrams by directed line segments. The length of the line represents the magnitude of the vector. The direction is represented by an arrow.

A vector is named either by using the letters at either end of the directed line segment $\overrightarrow{AB}, \overrightarrow{PQ}$, or by using a bold letter **u**, **v** and so on.

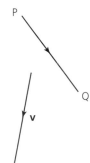

Components

Vectors may also be represented by their components.

$$\overrightarrow{AB} = \begin{pmatrix} 4 \\ 2 \end{pmatrix}, \ \overrightarrow{PQ} = \begin{pmatrix} 3 \\ -4 \end{pmatrix}, \ \mathbf{u} = \begin{pmatrix} -2 \\ 3 \end{pmatrix}, \ \mathbf{v} = \begin{pmatrix} -1 \\ -5 \end{pmatrix}$$

These are known as column vectors.

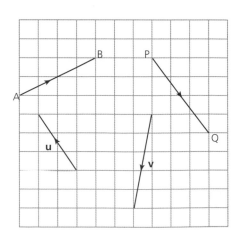

Magnitude

The magnitude (or length) of a vector is written $|\overrightarrow{AB}|$ or $|\mathbf{u}|$.

If $\mathbf{u} = \begin{pmatrix} x \\ y \end{pmatrix}$ then $|\mathbf{u}| = \sqrt{x^2 + y^2}$.

Example

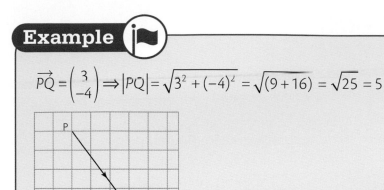

$$\overrightarrow{PQ} = \begin{pmatrix} 3 \\ -4 \end{pmatrix} \Rightarrow |\overrightarrow{PQ}| = \sqrt{3^2 + (-4)^2} = \sqrt{(9+16)} = \sqrt{25} = 5$$

Equal vectors and negative vectors

Vectors with the same magnitude and direction are equal. In the following diagram:

$$\mathbf{u} = \mathbf{v} = \begin{pmatrix} 2 \\ 3 \end{pmatrix}.$$

Vectors $\mathbf{u}$ and $\mathbf{w}$ have the same magnitude but opposite directions.

$$\mathbf{u} = -\mathbf{w}$$

$\overrightarrow{BA}$ and $\overrightarrow{AB}$ have the same magnitude but opposite directions. We say
$\overrightarrow{BA} = -\overrightarrow{AB}$.

For example, $\overrightarrow{AB} = \begin{pmatrix} 4 \\ -1 \end{pmatrix} \Rightarrow \overrightarrow{BA} = \begin{pmatrix} -4 \\ 1 \end{pmatrix}$.

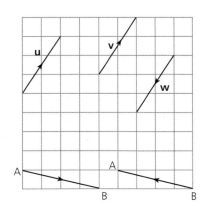

Adding and subtracting vectors

Vectors are added by placing them nose to tail.

$\overrightarrow{AB} + \overrightarrow{BC} = \overrightarrow{AC}$

$\overrightarrow{AC}$ is called the resultant vector.

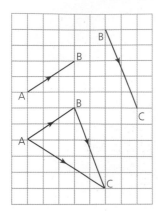

For example, $\overrightarrow{AC} = \overrightarrow{AB} + \overrightarrow{BC} = \begin{pmatrix} 3 \\ 2 \end{pmatrix} + \begin{pmatrix} 2 \\ -5 \end{pmatrix} = \begin{pmatrix} 5 \\ -3 \end{pmatrix}$

To subtract **v** from **u**, add −**v** to **u**.

So **u** − **v** = **u** + (−**v**)

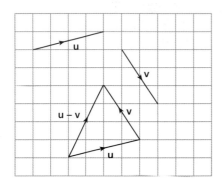

For example, $\mathbf{u} - \mathbf{v} = \begin{pmatrix} 4 \\ 1 \end{pmatrix} - \begin{pmatrix} 2 \\ -3 \end{pmatrix} = \begin{pmatrix} 2 \\ 4 \end{pmatrix}$

Example

An aeroplane sets off flying due east at 400 mph.
A wind blowing from the north at 50 mph blows the plane off course.
   a) Draw a diagram to show the resultant velocity of the plane.
   b) Calculate the resultant speed and bearing of the plane's course.

Solution

a)

speed^2 = 400^2 + 50^2 = 162 500 $\Rightarrow$ speed = $\sqrt{162\,500}$ = 403 mph
(to nearest mph)
b) $\tan a° = \frac{50}{400} \Rightarrow a° = \tan^{-1}\left(\frac{50}{400}\right) = 7°$ (to nearest degree)
$\Rightarrow$ bearing = 90° + 7° = 097°

Example

For this diagram, express in terms of **a** and **b**:

a) $\overrightarrow{PR}$

b) $\overrightarrow{SQ}$

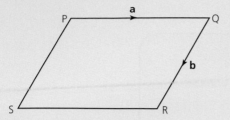

Solution

a) $\overrightarrow{PR} = \overrightarrow{PQ} + \overrightarrow{QR} = \mathbf{a} + \mathbf{b}$

b) $\overrightarrow{SQ} = \overrightarrow{SR} + \overrightarrow{RQ} = \mathbf{a} + (-\mathbf{b}) = \mathbf{a} - \mathbf{b}$

Three dimensions

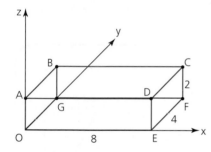

A is the point (0, 0, 2), B is (0, 4, 2), C is (8, 4, 2), D is (8, 0, 2), E is (8, 0, 0), F is (8, 4, 0) and G is (0, 4, 0).

$$\overrightarrow{OE} = \begin{pmatrix} 8 \\ 0 \\ 0 \end{pmatrix}, \quad \overrightarrow{OF} = \begin{pmatrix} 8 \\ 4 \\ 0 \end{pmatrix}, \quad \overrightarrow{OC} = \begin{pmatrix} 8 \\ 4 \\ 2 \end{pmatrix}$$

$$\overrightarrow{AC} = \overrightarrow{AD} + \overrightarrow{DC}$$

$$= \begin{pmatrix} 8 \\ 0 \\ 0 \end{pmatrix} + \begin{pmatrix} 0 \\ 4 \\ 0 \end{pmatrix} = \begin{pmatrix} 8 \\ 4 \\ 0 \end{pmatrix} \quad (note: \overrightarrow{AD} = \overrightarrow{OE}, \overrightarrow{DC} = \overrightarrow{OG})$$

$$\overrightarrow{DG} = \overrightarrow{DC} + \overrightarrow{CB} + \overrightarrow{BG}$$

$$= \begin{pmatrix} 0 \\ 4 \\ 0 \end{pmatrix} + \begin{pmatrix} -8 \\ 0 \\ 0 \end{pmatrix} + \begin{pmatrix} 0 \\ 0 \\ -2 \end{pmatrix} = \begin{pmatrix} -8 \\ 4 \\ -2 \end{pmatrix} \quad (note: \overrightarrow{CB} = -\overrightarrow{OE}, \overrightarrow{BG} = -\overrightarrow{OA})$$

Key points

* $\mathbf{a} + \mathbf{a} = 2\mathbf{a}$,
  $\mathbf{a} + \mathbf{a} + \mathbf{a} = 3\mathbf{a}$, etc.

a is parallel to 2**a**, 3**a**, etc.

In general k**a** is parallel to **a**.

$$2\mathbf{a} = 2\begin{pmatrix} 2 \\ 1 \end{pmatrix} = \begin{pmatrix} 4 \\ 2 \end{pmatrix},$$

$$3\mathbf{a} = 3\begin{pmatrix} 2 \\ 1 \end{pmatrix} = \begin{pmatrix} 6 \\ 3 \end{pmatrix} \text{ etc.}$$

$$\mathbf{a} - \mathbf{a} = \begin{pmatrix} 2 \\ 1 \end{pmatrix} - \begin{pmatrix} 2 \\ 1 \end{pmatrix} = \begin{pmatrix} 0 \\ 0 \end{pmatrix}$$

$\begin{pmatrix} 0 \\ 0 \end{pmatrix}$ is known as the zero vector.

Magnitude in three dimensions

If $\mathbf{u} = \begin{pmatrix} x \\ y \\ z \end{pmatrix}$, then $|\mathbf{u}| = \sqrt{(x^2 + y^2 + z^2)}$

Example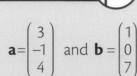

$\mathbf{a} = \begin{pmatrix} 3 \\ -1 \\ 4 \end{pmatrix}$ and $\mathbf{b} = \begin{pmatrix} 1 \\ 0 \\ 7 \end{pmatrix}$

Calculate the magnitude of $2\mathbf{a} - \mathbf{b}$.

Solution

$2\mathbf{a} - \mathbf{b} = 2\begin{pmatrix} 3 \\ -1 \\ 4 \end{pmatrix} - \begin{pmatrix} 1 \\ 0 \\ 7 \end{pmatrix} = \begin{pmatrix} 6 \\ -2 \\ 8 \end{pmatrix} - \begin{pmatrix} 1 \\ 0 \\ 7 \end{pmatrix} = \begin{pmatrix} 5 \\ -2 \\ 1 \end{pmatrix}$

$\Rightarrow |2\mathbf{a} - \mathbf{b}| = \sqrt{5^2 + (-2)^2 + 1^2} = \sqrt{(25 + 4 + 1)} = \sqrt{30}$

For practice

(The answers to the following questions are given in Appendix 1.)

1 Write down the components of
   a) $\overrightarrow{AB}$
   b) $\overrightarrow{PQ}$
   c) $\mathbf{u}$

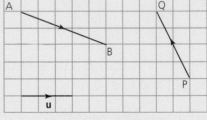

2 PQRSTU is a regular hexagon.
   Find a vector equal to:
   a) $\overrightarrow{PQ} + \overrightarrow{QR} + \overrightarrow{RS}$
   b) $\overrightarrow{UP} + \overrightarrow{TS}$
   c) $\overrightarrow{TU} - \overrightarrow{PU}$

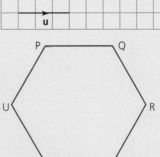

3 Use the diagram to name a vector equal to:
   a) $\mathbf{a} + 2\mathbf{b}$
   b) $\mathbf{c} - \mathbf{b}$
   c) $\mathbf{a} - \mathbf{d}$

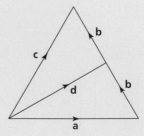

4 A boat sets off north east at 7·5 km/h but meets a current of
   4 km/h from the north west.
   a) Draw a diagram to show the resultant velocity of the boat.
   b) Calculate the boat's resultant speed and its bearing.

5 OABCD is a rectangular based pyramid of height 7 units.

The point D is vertically above the point of intersection of the diagonals of rectangle OABC.

State the coordinates of the points A, B, C and D.

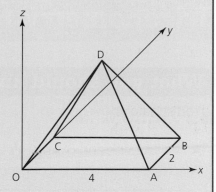

6 The diagram shows cuboid PQRSTUVW.

State the components of:

a) $\overrightarrow{PR}$

b) $\overrightarrow{SV}$

c) $\overrightarrow{VQ}$

d) $\overrightarrow{TR}$

e) $\overrightarrow{QW}$

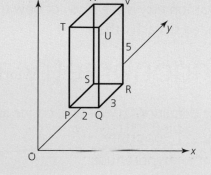

7 If $\mathbf{u} = \begin{pmatrix} -2 \\ 3 \\ 5 \end{pmatrix}$ and $\mathbf{v} = \begin{pmatrix} 1 \\ -2 \\ 4 \end{pmatrix}$, express in component form:

a) $\mathbf{u} + \mathbf{v}$

b) $\mathbf{u} - \mathbf{v}$

c) $2\mathbf{u} + 3\mathbf{v}$

8 Calculate the magnitude of each of these vectors.

a) $\overrightarrow{AB} = \begin{pmatrix} -4 \\ 8 \end{pmatrix}$

b) $\mathbf{w} = \begin{pmatrix} 6 \\ -2 \\ -3 \end{pmatrix}$

Trigonometry 1 – Triangles

What you should know

You should know how to:
★ calculate the area of a triangle using trigonometry
★ use the sine and cosine rules to find a side or angle in a triangle
★ use bearings with trigonometry to find a distance or direction.

Trigonometry is the study of how the sides and angles of a triangle are related to each other.

Trigonometry and right-angled triangles

The definitions of the sine, cosine and tangent ratios are:

$\sin x° = \dfrac{\text{opposite}}{\text{hypotenuse}}$

$\cos x° = \dfrac{\text{adjacent}}{\text{hypotenuse}}$

$\tan x° = \dfrac{\text{opposite}}{\text{adjacent}}$

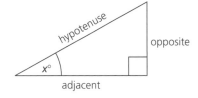

Use the mnemonic SOHCAHTOA to help you remember these definitions.

Example

The angle of elevation of the top of a building from a point 20 metres from the foot of the building is 55°.

Calculate the height of the building

Give the answer correct to 1 decimal place.

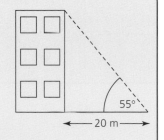

Solution

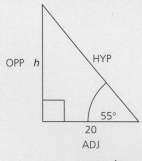

$\tan 55° = \dfrac{\text{opposite}}{\text{adjacent}} = \dfrac{h}{20}$

$\Rightarrow h = 20 × \tan 55° = 20 × 1·428... = 28·562... = 28·6 \text{ m (to 1 d.p.)}$

Example

The roof of this hut slopes at an angle of 20° to the horizontal.

The hut is 3 metres wide.

Calculate the length of the roof.

Give the answer correct to 2 decimal places.

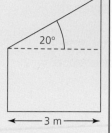

Solution

$\cos 20° = \dfrac{\text{adjacent}}{\text{hypotenuse}} = \dfrac{3}{r}$

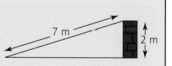

$\Rightarrow r \times \cos 20° = 3$

$\Rightarrow \quad r = \dfrac{3}{\cos 20°} = \dfrac{3}{0.939...} = 3.192... = 3.19 \text{ m (to 2 d.p.)}$

Example

The diagram shows a side view of a ramp.

Calculate the size of the angle between the ramp and the horizontal.

Give the answer correct to 1 decimal place.

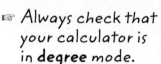

Solution

$\sin x° = \dfrac{\text{opposite}}{\text{hypotenuse}} = \dfrac{2}{7}$

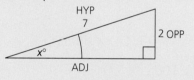

$\Rightarrow x° = \sin^{-1}\left(\dfrac{2}{7}\right) = \sin^{-1}(0.285...)$

$\quad = 16.601...° = 16.6° \text{ (to 1 d.p.)}$ — To evaluate $\sin^{-1}\left(\dfrac{2}{7}\right)$, use the 'Inv', '2nd fn' or 'shift' button on your calculator.

Remember

☞ Always check that your calculator is in **degree** mode.
☞ Do not round any values until the end of the calculations.

Trigonometry and any triangle

When a triangle is not right-angled, the three ratios can't be used as in the examples above. The methods to be used are described in the remainder of this chapter.

Area of a triangle

Example

Calculate the area of this triangle.

Give the answer correct to 3 significant figures.

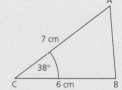

Solution

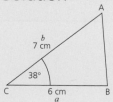

Area $= \dfrac{1}{2}ab\sin C$

$= \dfrac{1}{2} \times 6 \times 7 \times \sin 38°$

$= 12.928... = 12.9 \text{ cm}^2 \text{ (to 3 s.f.)}$

Key points

Area $= \dfrac{1}{2}ab\sin C$

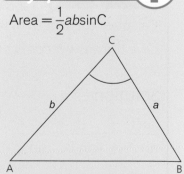

In order to use this formula you need to know the lengths of two sides and their included angle.

Example

Calculate the area of parallelogram PQRS.

Give the answer correct to 3 significant figures.

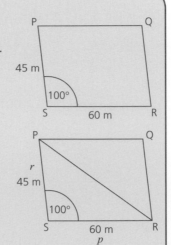

Solution

Area of triangle PSR = $\frac{1}{2}prsinS$

$= \frac{1}{2} \times 60 \times 45 \times sin\,100°$

$= 1329·49...$

$\Rightarrow$ area of parallelogram PQRS

$= 2 \times 1329·49...$

$= 2658·980...$

$= 2660\,m^2$ (to 3 s.f.)

Example

The area of triangle CDE is 12 cm^2.

Calculate the length of CD to the nearest centimetre.

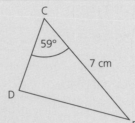

Solution

Area of triangle CDE = $\frac{1}{2}desinC \Rightarrow 12 = \frac{1}{2} \times 7 \times CD \times sin\,59°$

$\Rightarrow 12 = 3·00... \times CD$

$\Rightarrow CD = 12 \div 3·00...$

$= 4\,cm$ (to nearest cm)

The sine rule

Example

Calculate the length of BC in triangle ABC.

Give the answer correct to one decimal place.

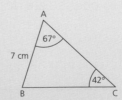

Solution

$\frac{a?}{sin\,A✓} = \frac{b}{sin\,B} = \frac{c✓}{sin\,C✓}$

Put a '?' next to the part you want to find, and a tick next to parts that you were given.

$\frac{a}{sin\,67°} = \frac{7}{sin\,42°}$

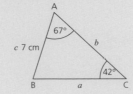

$\Rightarrow a = \frac{7 \times sin\,67°}{sin\,42°} = 9·629... = 9·6\,cm$ (to 1 d.p.)

Key points !

In triangle ABC:

$\frac{a}{sin\,A} = \frac{b}{sin\,B} = \frac{c}{sin\,C}$

Example

Calculate the size of angle PQR in triangle PQR.

Give the answer correct to one decimal place.

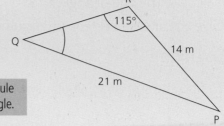

Solution

$$\frac{\sin P}{p} = \frac{\sin Q?}{q✓} = \frac{\sin R ✓}{r✓}$$ ⎯ It is easier to use this form of the sine rule when you are finding the size of an angle.

$$\frac{\sin Q}{14} = \frac{\sin 115°}{21}$$

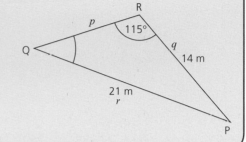

$$\Rightarrow \sin Q = \frac{14 \times \sin 115°}{21} = 0·604…$$

$$\Rightarrow Q = \sin^{-1}0·604… = 37·171… = 37·2° \text{ (to 1 d.p.)}$$

The cosine rule

Key points

In triangle ABC:

$$a^2 = b^2 + c^2 - 2bc\cos A$$

or

$$\cos A = \frac{b^2 + c^2 - a^2}{2bc}$$

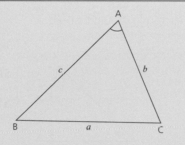

Example

Calculate the length of GH.

Give the answer correct to 3 significant figures.

Solution

To find a length use the first form of the cosine rule:

$$f^2 = g^2 + h^2 - 2gh\cos F$$
$$= 11^2 + 8^2 - 2 \times 11 \times 8 \times \cos70°$$ ⎯ Enter **all** of this into your calculator.
$$= 124·804…$$
$$\Rightarrow f = \sqrt{124·804}…$$
$$= 11·171…$$
$$= 11·2 \text{ cm (to 3 s.f.)}$$

79

Example

Calculate the size of angle XYZ.

Give the answer correct to 3 significant figures.

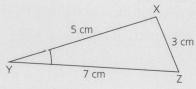

Solution

To find the size of an angle use the second form of the cosine rule:

$$\cos Y = \frac{x^2 + z^2 - y^2}{2xz}$$

$$= \frac{7^2 + 5^2 - 3^2}{2 \times 7 \times 5}$$

$$= \frac{65}{70}$$

$$\Rightarrow Y = \cos^{-1}\left(\tfrac{65}{70}\right) = 21{\cdot}786\ldots = 21{\cdot}8° \text{ (to 3 s.f.)}$$

Key points

The area formula, the sine rule and both versions of the cosine rule will be given in the formula list for the National 5 exam. The table below gives a guide as to when each rule should be used.

What you are given		What you should use
A side and the angle opposite this side a) **side** required b) **angle** required		**Sine rule** a) $\dfrac{a}{\sin A} = \dfrac{b}{\sin B} = \dfrac{c}{\sin C}$ b) $\dfrac{\sin A}{a} = \dfrac{\sin B}{b} = \dfrac{\sin C}{c}$
Two sides and the included angle a) **side** required b) **area** required		a) **Cosine rule** $a^2 = b^2 + c^2 - 2bc\cos A$ b) **Area formula** Area $= \tfrac{1}{2}ab\sin C$
All three sides **angle** required		**Cosine rule** $\cos A = \dfrac{b^2 + c^2 - a^2}{2bc}$

Problems involving sine rule, cosine rule and area formula

Example

Two planes leave from airport A at the same time.
Plane B flies on a bearing of 065° at 400 mph.
Plane C flies on a bearing of 143° at 500 mph.

How far apart are the two planes after one hour?
Give the answer correct to 2 significant figures.

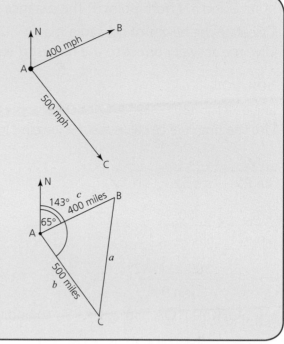

Solution

angle BAC = 143° − 65° = 78°

$a^2 = b^2 + c^2 - 2bc\cos A$

$ = 400^2 + 500^2 - 2 \times 400 \times 500 \times \cos 78°$

$ = 326\,835 \cdot 323...$

$\Rightarrow a = \sqrt{326\,835 \cdot 323...}$

$ = 571 \cdot 695...$

$ = 570$ miles (to 2 s.f.)

Example

Three oil rigs, P, Q and R are situated in a desert.
P is 150 km due north of Q, and 290 km from R.
R is on a bearing of 120° from Q.
Calculate the bearing of R from P.
Give the answer correct to the nearest degree.

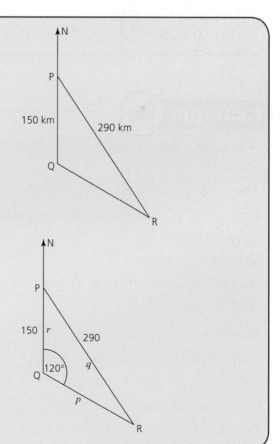

Solution

Use the sine rule to calculate the size of angle PRQ.

$$\frac{\sin P}{p} = \frac{\sin Q✔}{q✔} = \frac{\sin R?}{r✔}$$

$$\frac{\sin R}{150} = \frac{\sin 120°}{290}$$

$$\Rightarrow \sin R = \frac{150 \times \sin 120°}{290} = 0 \cdot 447...$$

$\Rightarrow R = \sin^{-1} 0 \cdot 447... = 26 \cdot 611... = 27°$ (to nearest degree)

$\Rightarrow$ angle QPR = 180° − 120° − 27° = 33°

$\Rightarrow$ bearing of R from P = 180° − 33° = 147°.

Example

The angle of elevation to the top of a radio mast, R, is 39° from
point S and 47° from point T. The distance from S to T is 100 metres.

Calculate the height of the radio mast.
Give the answer correct to the nearest metre.

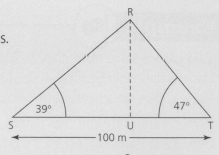

Solution

angle SRT = 180° − 39° − 47° = 94°.
Use the sine rule to calculate the length of RS in triangle RST.

$$\frac{r✓}{\sin R✓} = \frac{s}{\sin S✓} = \frac{t?}{\sin T✓}$$

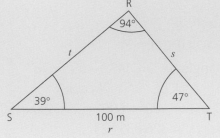

$$\frac{t}{\sin 47°} = \frac{100}{\sin 94°}$$

$$\Rightarrow \quad t = \frac{100 \times \sin 47°}{\sin 94°} = 73.313\ldots = 73.3 \text{ m (to 1 d.p.)}$$

Use SOHCAHTOA in triangle RSU to find the height of the mast.

$$\sin 39° = \frac{\text{opposite}}{\text{hypotenuse}} = \frac{RU}{73.313\ldots}$$

$\Rightarrow$ RU = 73.313 × sin 39° = 46.137…
$\Rightarrow$ height of radio mast = 46 m (to nearest metre)

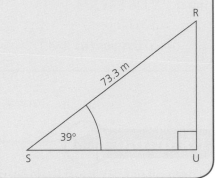

Example

A gift box, 7 centimetres high, is prism
shaped.

The cross-section is a regular hexagon.

Each vertex of the hexagon is 5 centimetres
from the centre C.

Calculate the volume of the box.

Give the answer correct to 3 significant
figures.

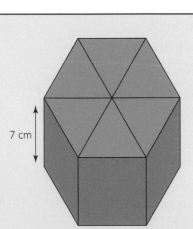
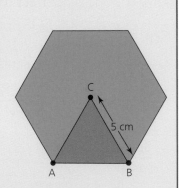

Solution

Angle ACB = 360° ÷ 6 = 60°.
area of triangle ACB = $\frac{1}{2} ab \sin C = \frac{1}{2} \times 5 \times 5 \times \sin 60° = 10.825\ldots$ cm^2
$\Rightarrow$ area of hexagon = 6 × 10.825… = 64.951…cm^2
$\Rightarrow$ volume of prism = Ah = 64.951… × 7 = 454.663… = 455cm^3 (to 3 s.f.)

For practice

(The answers to the following questions are given in Appendix 1.)

1 The diagram shows triangle DFE.
Calculate
a) the area of triangle DFE
b) the length of EF. Give both answers correct to 3 significant figures.

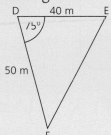

2 The diagram shows triangle ABC.
Calculate the length of BC.
Give the answer correct to 1 decimal place.

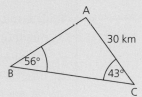

3 The diagram shows triangle STU.
Calculate the size of angle TUS.
Give the answer to the nearest degree.

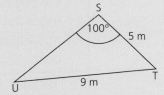

4 The area of triangle PQR is 17 cm².
Calculate the size of angle QPR.
Give the answer to 3 significant figures.

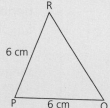

5 Calculate the size of angle KLM.
Give the answer correct to 1 decimal place.

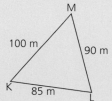

6 A ship sails on a bearing of 055° from port X to a port Z, 80 kilometres away.
Port Y is due east of port X.
Another ship sails on a bearing of 290° from port Y to port Z.
Calculate the distance between port X and port Y.
Give your answer to the nearest kilometre.

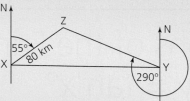

7 An aeroplane flies from its base on a bearing of 030° for 500 kilometres.
It then changes direction and flies on a bearing of 160° for a further 750 kilometres.
The plane then returns directly to its base.
Calculate the distance and bearing of the last leg of its flight.
Give both answers to the nearest whole number.

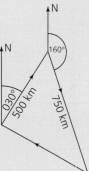

8 The diagram shows two buildings of equal height.
P and Q represent points on the top of the buildings and R represents a point on the ground between them.
From P, the angle of depression of R is 39°.
From Q, the angle of depression of R is 47°.
The distance PQ is 40 metres.
Calculate the height of the buildings.
Give the answer correct to 3 significant figures.

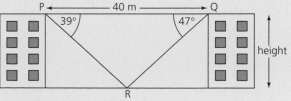

What you should know

You should know how to:
- ★ work with the graphs of trigonometric functions
- ★ work with the sine, cosine and tangent of angles from 0° up to 360°
- ★ solve basic trigonometric equations
- ★ work with two important trigonometric identities.

Graphs of the Functions $y = \sin x°$, $y = \cos x°$ and $y = \tan x°$

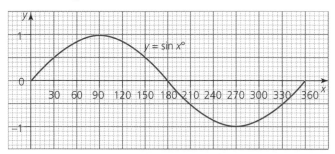

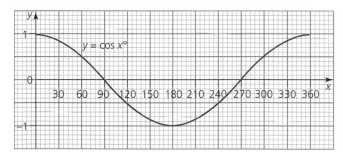

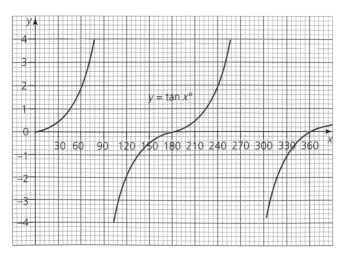

Key points

- ✳ The maximum and minimum values of the sine and cosine functions are ±1. We say that the **amplitude** of these functions is 1. The amplitude is the distance from the midpoint to the highest and lowest points of the graph.

- ✳ The sine and cosine functions repeat themselves every 360°. We say that the **period** of these functions is 360°.

- ✳ The period of the tangent function is 180°.

Graphs of related trigonometric functions

Key points

function	amplitude	maximum	minimum	period
$y = a \sin bx°$	a	a	$-a$	$\dfrac{360°}{b}$
$y = a \cos bx°$	a	a	$-a$	$\dfrac{360°}{b}$
$y = a \tan bx°$	–	–	–	$\dfrac{180°}{b}$

For $y = f(x)$ where $f(x) = \sin x°$, $f(x) = \cos x°$ or $f(x) = \tan x°$:

$y = f(x) + c$	If c is positive the graph of f(x) moves c units up. If c is negative the graph of f(x) moves c units down.
$y = f(x - d)$	If d is positive the graph of f(x) moves d units right. If d is negative the graph of f(x) moves c units left. d is known as the phase angle.

Example

Write down the equations of the following graphs.

a)

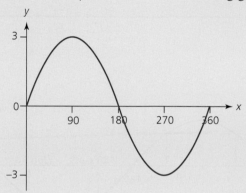

b)

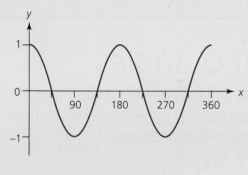

Solution

a) This is a sine function.
Its amplitude is 3.
Its period is 360°.
Hence $y = 3 \sin x°$.

b) This is a cosine function.
Its amplitude is 1.
Its period is 180°. Hence $y = \cos 2x°$.

⇨

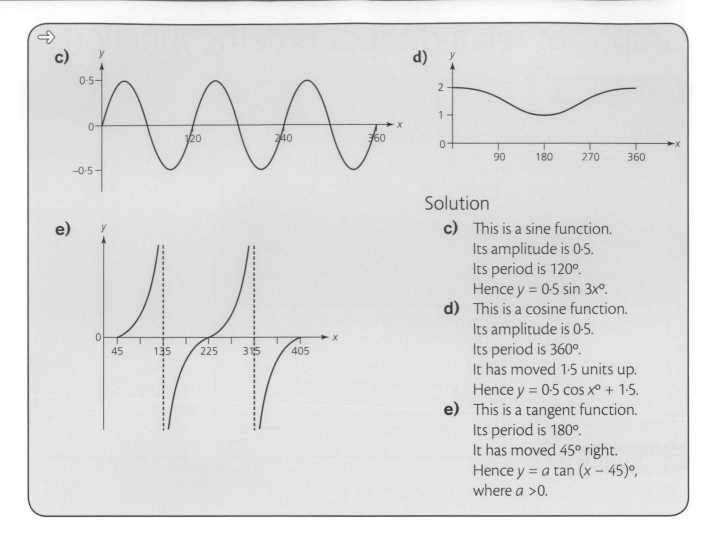

c)

d)

Solution

c) This is a sine function.
Its amplitude is 0·5.
Its period is 120°.
Hence $y = 0·5 \sin 3x°$.

d) This is a cosine function.
Its amplitude is 0·5.
Its period is 360°.
It has moved 1·5 units up.
Hence $y = 0·5 \cos x° + 1·5$.

e) This is a tangent function.
Its period is 180°.
It has moved 45° right.
Hence $y = a \tan (x - 45)°$,
where $a > 0$.

Related angles

The symmetry of the graphs of trigonometric functions can be used to find angles which have the same sine, cosine or tangent.

This graph shows that

- $\sin (180 - a)° = \sin a°$
- $\sin (180 + a)° = \sin (360 - a)° = - \sin a°$.

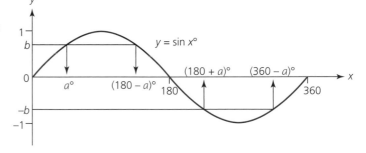

This graph shows that

- $\cos (360 - a)° = \cos a°$
- $\cos (180 - a)° = \cos (180 + a)° = - \cos a°$.

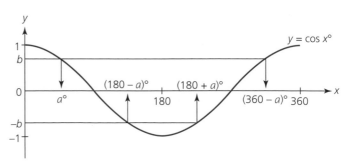

This graph shows that
- $\tan (180 + a)° = \tan a°$
- $\tan (180 - a)° = \tan (360 - a)° = - \tan a°$.

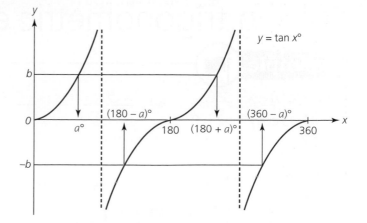

Key points

The above relationships are summarised in the four-quadrant diagram shown below.

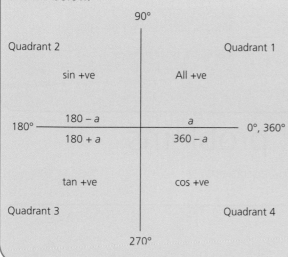

Example

Express the following in terms of the sine, cosine or tangent of an acute angle:

    a) $\sin 150°$

    b) $\cos 220°$

    c) $\tan 325°$.

Solution

    a) $\sin 150° = \sin (180 - 30)° = \sin 30°$

    b) $\cos 220° = \cos (180 + 40)° = - \cos 40°$

    c) $\tan 325° = \tan (360 - 35)° = - \tan 35°$

Solving trigonometric equations

Example

Solve for $0 \leq x < 360$.

a) $3\cos x° - 2 = 0$

b) $4\tan x° + 9 = 2$

Solution

a) $3\cos x° - 2 = 0 \Rightarrow 3\cos x° = 2$

$\Rightarrow \cos x° = \frac{2}{3}$

$\cos x°$ is positive, so x is in quadrants 1 and 4 and

$\cos^{-1}\left(\frac{2}{3}\right) = 48.2°$ (to 1 d.p.)

$\Rightarrow \quad x = 48.2$ and $(360 - 48.2)$

$\Rightarrow \quad x = 48.2$ and 311.8

b) $4\tan x° + 9 = 2 \Rightarrow 4\tan x° = -7$

$\Rightarrow \tan x° = \left(\frac{-7}{4}\right)$

$\tan x°$ is negative, so x is in quadrants 2 and 4 and

$\tan^{-1}\left(\frac{7}{4}\right) = 60.3°$ (to 1 d.p.)

Beware, put $\tan^{-1}\left(\frac{7}{4}\right)$ into your calculator, **not** $\tan^{-1}\left(\frac{-7}{4}\right)$.

$\Rightarrow \quad x = (180 - 60.3)$ and $(360 - 60.3)$

$\Rightarrow \quad x = 119.7$ and 299.7

Trigonometric equations: problems

You may be asked to solve problems involving trigonometric equations.

Example

A ferris wheel rotates at a steady rate.

The height, h metres, of a point A above the ground at time t seconds is given by the equation $h = 25 + 20\sin t°$.

a) Calculate the height of point A at time 30 seconds.

b) Find two times during the first turn of the wheel when point A is 13 metres above the ground.

c) Find the maximum height of point A above the ground.

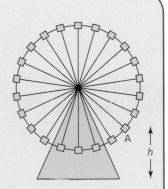

Solution

a) $t = 30 \Rightarrow h = 25 + 20 \times \sin 30° = 25 + 10 = 35$ metres.

b) $h = 13 \Rightarrow 13 = 25 + 20\sin t°$

$\Rightarrow 20\sin t° = -12$

$\Rightarrow \sin t° = \frac{-12}{20} = \frac{-3}{5}$

$\sin t°$ is negative, so t is in quadrants 3 and 4 and $\sin^{-1}\left(\frac{3}{5}\right) = 37°$ (to nearest degree)

Beware, put $\sin^{-1}\left(\frac{3}{5}\right)$ into your calculator, **not** $\sin^{-1}\left(\frac{-3}{5}\right)$.

$\Rightarrow t = (180 + 37)$ and $(360 - 37)$

$\Rightarrow t = 217$ and 323

Hence point A is 13 metres above the ground after 217 seconds and 323 seconds.

c) maximum value of $\sin t°$ is 1

$\Rightarrow$ maximum value of $h = 25 + 20\sin t°$ is $h = 25 + 20 \times 1 = 25 + 20 = 45$ hence the maximum height of point A above the ground is 45 metres.

Trigonometric identities

An **identity** involving x is a formula that is true for all values of x.

Key points !

There are two important trigonometric identities which you are expected to know and be able to work with. These two formulae are true for all values of x.

$$\tan x^\circ = \frac{\sin x^\circ}{\cos x^\circ} \text{ and } \sin^2 x^\circ + \cos^2 x^\circ = 1$$

Note: $\sin^2 x^\circ + \cos^2 x^\circ = 1$ can be rearranged to give

✳ $\sin^2 x^\circ = 1 - \cos^2 x^\circ$

✳ $\cos^2 x^\circ = 1 - \sin^2 x^\circ$

Example

Prove that $(\sin x^\circ + \cos x^\circ)^2 = 1 + 2 \sin x^\circ \cos x^\circ$.

Solution

$$
\begin{aligned}
\text{Left Hand Side (LHS)} &= (\sin x^\circ + \cos x^\circ)^2 \\
&= (\sin x^\circ + \cos x^\circ)(\sin x^\circ + \cos x^\circ) \\
&= \sin^2 x^\circ + \sin x^\circ \cos x^\circ + \cos x^\circ \sin x^\circ + \cos^2 x^\circ \\
&= \sin^2 x^\circ + \cos^2 x^\circ + 2\sin x^\circ \cos x^\circ \\
&= 1 + 2 \sin x^\circ \cos x^\circ \\
&= \text{Right Hand Side (RHS)}
\end{aligned}
$$

Example

If $\cos x^\circ = \frac{1}{2}$, calculate the exact value of $\sin x^\circ$, $0 < x < 90$.

Solution

$$
\begin{aligned}
\sin^2 x^\circ + \cos^2 x^\circ = 1 \Rightarrow \sin^2 x^\circ &= 1 - \cos^2 x^\circ \\
&= 1 - \left(\tfrac{1}{2}\right)^2 \\
&= 1 - \tfrac{1}{4} \\
&= \tfrac{3}{4} \\
\Rightarrow \sin x^\circ &= \sqrt{\tfrac{3}{4}} \\
&= \frac{\sqrt{3}}{2}
\end{aligned}
$$

Example

Simplify $\tan x^\circ \cos x^\circ$

Solution

$$
\begin{aligned}
\tan x^\circ \cos x^\circ &= \frac{\sin x^\circ}{\cos x^\circ} \times \cos x^\circ \\
&= \sin x^\circ
\end{aligned}
$$

For practice

(The answers to the following questions are given in Appendix 1)

1 Write down the equations of the following graphs.

a)

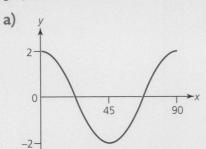

b)

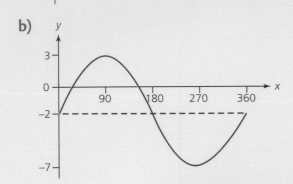

c)

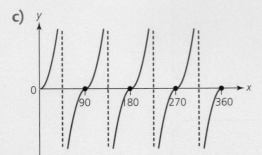

d)

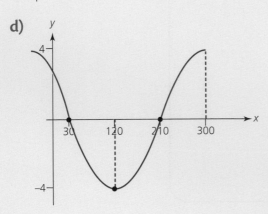

2 Sketch the graphs of the following functions for $0 \leq x \leq 360$.

a) $y = 3 \sin 2x°$

b) $y = 0.5 \cos x° + 1$

3 The diagram shows the graph with equation $y = 5 \sin x° + 7$.

Write down the coordinates of the points A and B.

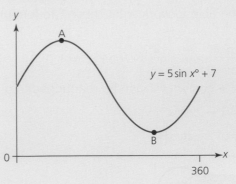

4 Solve for $0 \leq x < 360$. Give the answers correct to one decimal place.

a) $4 \sin x° - 1 = 0$

b) $5 \tan x° + 3 = 1$

c) $8 \cos x° + 11 = 6$

5 The diagram shows the graph with equation $y = 1 - 3 \cos x°$.

The points $(60, p)$ and $(q, 2)$ lie on the graph as shown.

Calculate the values of p and q.

Give the answers to one decimal place.

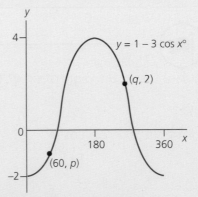

6 If $\sin x° = \frac{12}{13}$ and $0 \leq x \leq 90$, find the exact values of:

a) $\cos x°$

b) $\tan x°$

7 Simplify $\dfrac{\sqrt{1 - \cos^2 x°}}{\cos x°}$.

8 Prove that $\cos x°(\sin x° + \cos x°) + \sin x° (\sin x° - \cos x°) = 1$.

You should know how to:
★ compare data sets using calculated/determined statistics:
  – semi-interquartile range
  – standard deviation
★ determine the equation of a best-fitting straight line on a scattergraph and use it to estimate y given x.

When comparing data sets it is useful to calculate averages and measures of spread.

Averages

There are three main types of average: mean, median and mode.

$$\text{mean} = \frac{\text{sum of statistics}}{\text{number of statistics}}$$

median = the **middle** statistic, when the statistics are in **ascending order**

mode = the statistic which occurs **most often**

Example

Find the mean, median and mode of the following data sets:
a) 11 8 10 7 11
b) 21 40 25 28 21 36.

Solution
a) $\text{mean} = \frac{11 + 8 + 10 + 7 + 11}{5} = \frac{47}{5} = 9.4$

median: first arrange the statistics in ascending order:
        7 8 <u>10</u> 11 11
hence median (middle statistic) = 10
mode: 11 (occurs most)

b) $\text{mean} = \frac{21 + 40 + 25 + 28 + 21 + 36}{6} = \frac{171}{6} = 28.5$

median: first arrange the statistics in ascending order:

21 21 <u>25</u> <u>28</u> 36 40 —— There is no middle number, so calculate the mean of the numbers on either side of the middle.

hence median $= \frac{25 + 28}{2} = 26.5$

mode: 21 (occurs most)

Measures of spread

There are three measures of spread which you will be expected to use:

range, semi-interquartile range and standard deviation.

Range

range = largest statistic − smallest statistic

> **Example**
>
> Find the range of the following data set:
>
> 12   6   8   12   5   27   9   16
>
> ### Solution
> largest number = 27
>
> smallest number = 5
>
> hence range = 27 − 5 = 22.
>
> Note: The range measures the spread of all the data. It is influenced by extreme high or low values of data and can be misleading.

Quartiles, interquartile range and semi-interquartile range

Quartiles divide the data into four equal sized groups just as the median divides the data into two equal sized groups.

> **Example**
>
> Find the quartiles for each of the following data sets:
> - **a)** 12   23   19   42   38   27
> - **b)** 5   9   2   6   12   8   11   14   3
>
> ### Solution
> **a)** first arrange the statistics in order:
>
> (12 <u>19</u> 23) (27 <u>38</u> 42)
>
> $\Rightarrow$ median, $Q_2 = \frac{23 + 27}{2} = \frac{50}{2} = 25$
>
> lower quartile, $Q_1 = 19$
> upper quartile, $Q_3 = 38$
>
> **b)** first arrange the statistics in order:
>
> (2 <u>3</u> <u>5</u> 6) 8 (9 <u>11</u> <u>12</u> 14)
>
> $\Rightarrow Q_2 = 8$
>
> $Q_1 = \frac{3 + 5}{2} = \frac{8}{2} = 4$
>
> $Q_3 = \frac{11 + 12}{2} = \frac{23}{2} = 11\cdot5$

> ### *Key points*
>
> To find the quartiles:
> * arrange the statistics in ascending order
> * find the median – this is the *second quartile*, Q_2
> * find the middle statistic for the lower group (*the lower quartile Q_1*) and for the upper group (*the upper quartile Q_3*).

Key points

The **interquartile range** (IQR) measures the spread of the middle 50% of a data set.

Interquartile range = upper quartile − lower quartile i.e. IQR = $Q_3 - Q_1$.

The **semi-interquartile** range is half of the interquartile range i.e.

$$SIQR = \frac{Q_3 - Q_1}{2}$$

Example

Calculate the semi-interquartile range for this data set:

10, 14, 12, 20, 21, 16, 12, 13, 18, 19, 21, 22, 19, 16

Solution

first arrange the statistics in order:

(10 12 12 $\underline{13}$ 14 16 16) (18 19 19 $\underline{20}$ 21 21 22)

$\Rightarrow$ median $= \frac{16+18}{2} = \frac{34}{2} = 17$, $Q_1 = 13$ and $Q_3 = 20$

$\Rightarrow SIQR = \frac{Q_3 - Q_1}{2} = \frac{20-13}{2} = \frac{7}{2} = 3.5$

Key points

A **box plot** is a useful way of illustrating the quartiles along with the smallest and largest statistics in a data set.

This box plot illustrates the data set in the example above.

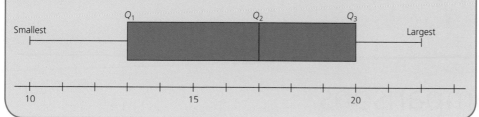

Standard deviation

Standard deviation is a more reliable measure of spread since **all** of the statistics are used in its calculation rather than just two in the case of calculating the range or interquartile range. Standard deviation measures the average deviation (difference) of each statistic from the mean. The larger the value of the standard deviation, the larger is the spread of the data from the mean.

Key points

The standard deviation can be calculated from either of the formulae:

$$s = \sqrt{\frac{\sum(x-\bar{x})^2}{n-1}} \text{ or } s = \sqrt{\frac{\sum x^2 - \frac{(\sum x)^2}{n}}{n-1}}, \text{ where } n \text{ is the sample size.}$$

Example

Calculate the standard deviation for the following data set: 6 8 9 9 11 14

Solution

Using the first formula:

mean $= \bar{x} = \dfrac{6 + 8 + 9 + 9 + 11 + 14}{6} = \dfrac{57}{6} = 9{\cdot}5$

x	$x - \bar{x}$	$(x - \bar{x})^2$
6	−3·5	12·25
8	−1·5	2·25
9	−0·5	0·25
9	−0·5	0·25
11	1·5	2·25
14	4·5	20·25
		$\Sigma(x - \bar{x})^2 = 37{\cdot}5$

$$s = \sqrt{\frac{\Sigma(x - \bar{x})^2}{n-1}}$$

$$= \sqrt{\frac{37{\cdot}5}{5}}$$

$$= \sqrt{7{\cdot}5}$$

$$= 2{\cdot}74 \ \text{(to 2 d.p.)}$$

Using the second formula: $s = \sqrt{\dfrac{\Sigma x^2 - \dfrac{(\Sigma x)^2}{n}}{n-1}}$

x	x^2
6	36
8	64
9	81
9	81
11	121
14	196
$\Sigma x = 57$	$\Sigma x^2 = 579$

$$= \sqrt{\frac{579 - \frac{57^2}{6}}{5}}$$

$$= \sqrt{\frac{579 - 541{\cdot}5}{5}}$$

$$= \sqrt{\frac{37{\cdot}5}{5}} = \sqrt{7{\cdot}5}$$

$$= 2{\cdot}74$$

Statistical comparisons

You are expected to be able to compare different sets of data.

Example

A group of students sit a maths test in December and another one in May.

The median mark and semi-interquartile range are as follows:

	Median	Semi-interquartile range
December	56%	11%
May	63%	7%

Make two valid comparisons between the two sets of data.

Solution

In May, the marks (on average) are better (since the median mark is bigger), and they are less varied (since the semi-interquartile range is smaller).

Example

The price, in pence per litre, of petrol at some city garages and some rural garages is recorded. The mean and standard deviation of the prices are as follows:

	Mean	Standard deviation
City garages	129·7	3·2
Rural garages	143·1	7·8

Make two valid comparisons between the two sets of data.

Solution

In the city garages, the price (on average) is lower (since the mean price is smaller) and it is more consistent (since the standard deviation is smaller).

Scattergraphs

A scattergraph is a collection of points on a coordinate diagram representing two sets of data. When there is a connection or correlation between the two sets of data a **line of best fit** can be drawn on the scattergraph. The line of best fit should best represent the trend of the data. There should be roughly the same number of points above and below the line.

Example

This scattergraph shows the marks scored in a French test by a group of students against their marks scored in a German test. The line of best fit has been drawn.

Use the line of best fit to estimate the mark scored in the German test by a student who scored 25 in the French test.

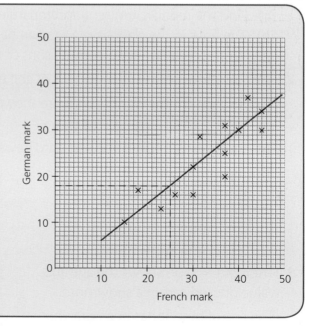

Solution

By projecting a French mark of 25 on to the line of best fit, you see that the student's estimated German mark is 18.

Hints & tips

In a question like this, you may be asked to find the equation of the line of best fit. Remember, the equation of a straight line through the point (a, b) with gradient m, is y − b = m(x − a).

Example

For the scattergraph on page 95:

   a) Find the equation of the line of best fit.

   b) Use the equation of the line of best fit to estimate the mark scored in the German test by a student who scored 25 in the French test.

Solution

   a) Two well-spaced points on the line, e.g. (15, 10) and (40, 30), can be used to calculate the gradient, and any point on the line, e.g. (15, 10) can be used for (a, b).

Hence $m = \dfrac{y_2 - y_1}{x_2 - x_1} = \dfrac{30 - 10}{40 - 15} = \dfrac{20}{25} = \dfrac{4}{5}$; $(a, b) = (15, 10)$.

The equation of the line of best fit is therefore $y - 10 = \dfrac{4}{5}(x - 15)$

$$\Leftrightarrow y - 10 = \tfrac{4}{5}x - 12$$

$$\Leftrightarrow \quad y = \tfrac{4}{5}x - 2$$

   b) Substitute $x = 25$ into the equation $y = \tfrac{4}{5}x - 2$

$$\Rightarrow y = \tfrac{4}{5} \times 25 - 2$$

$$= 20 - 2$$

$$= 18$$

Hence, the estimated mark for the student in the German test is 18.

For practice

(The answers to the following questions are given in Appendix 1.)

1 The heights, in centimetres, of the thirteen girls in a Primary 7 class are shown below.

   130 116 129 145 111 138 131 122 144 147 126 138 136

   a) Find the median and semi-interquartile range of the girls' heights.

   b) The boys in the class have a median height of 124cm and an interquartile range of 11 cm. Write down two valid comparisons of the girls' heights and the boys' heights.

2 Jack and Tom played eight rounds of golf each.

   Jack's scores for the eight rounds are shown below.

   72 72 71 68 69 71 73 72

   a) Calculate the mean and standard deviation of Jack's scores.

   b) Tom's scores have a mean of 73 and a standard deviation of 2·1. Write down two valid comparisons between Jack's scores and Tom's scores.

3 This scattergraph shows the age and value of used cars in a showroom.

   The line of best fit has been drawn.

   a) Find the equation of the line of best fit.

   b) Use the equation of the line of best fit to estimate the value of a nine-year-old used car.

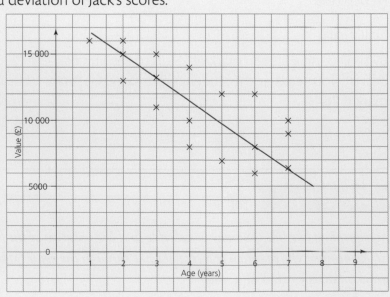

Chapter 1

1 a) 50 000
b) 0·067
c) 209 000
2 £2680·24
3 £92 610
4 £1800
5 £17 500

6 £460
7 $1\frac{5}{21}$
8 $5\frac{19}{20}$
9 $\frac{11}{24}$
10 $4\frac{3}{5}$
11 $\frac{14}{27}$

12 $4\frac{1}{6}$
13 $\frac{9}{10}$
14 $\frac{12}{35}$
15 $2\frac{2}{3}$
16 $1\frac{1}{4}$

Chapter 2

1 a) $2x^2 - 10x$
b) $y - 3y^2$
2 a) $5 - 7a$
b) $19b - 5$
c) $10c + 32$
d) $x^2 + 5x - 36$
e) $10y^2 + 51y + 27$
f) $n^2 - 16n + 64$
g) $4k^2 + 28k + 49$
h) $m^3 + 5m^2 + m - 15$
i) $8t^3 - 10t^2 + 13t - 5$

3 a) $6(4x - 3y)$
b) $5b(7a + 1)$
c) $2c(c + 5)$
d) $(m + 9)(m - 9)$
e) $(5 + n)(5 - n)$
f) $(3p + 10q)(3p - 10q)$
g) $(x + 2)(x - 5)$
h) $(y - 1)(y - 8)$
i) $(3s + 2)(s + 3)$

4 a) $2(x + 3)(x - 3)$
b) $5(d + 4)(d - 2)$
c) $3(2k + 3)(k + 1)$
5 a) $(x + 4)^2 - 6$
b) $(x - 7)^2 + 1$
c) $(x + 1)^2 - 2$
d) $(x - 2)^2 - 9$
e) $\left(x + \frac{3}{2}\right)^2 - \frac{1}{4}$

Chapter 3

1 $\frac{5}{6}$
2 a) 9
b) 125
c) 2
3 a) $2a^3$
b) b^2
c) $c^{\frac{3}{2}}$
d) $16d^{14}$
e) $50e^9$
4 $\frac{1}{a^4}$
5 a) $m + 2$

b) $n^2 - 1$
6 a) 1 230 000 000
b) 0·000 21
7 a) $6·3 \times 10^7$
b) 8×10^{-3}
8 a) $1·09 \times 10^{10}$
b) $3·8 \times 10^{-14}$
9 500 seconds
10 $1·92 \times 10^{16}$
11 a) $3\sqrt{2}$
b) $4\sqrt{3}$

c) $\frac{6}{7}$
d) $\frac{2}{5}$
12 a) $7\sqrt{3}$
b) $2\sqrt{5} + 6$
c) 1
d) $5 - 2\sqrt{6}$
13 a) $\frac{3\sqrt{5}}{5}$
b) $\frac{\sqrt{2}}{4}$
c) $\frac{3\sqrt{6}}{2}$
d) $\frac{3 + \sqrt{2}}{7}$

Chapter 4

1 a) $\frac{2}{3b}$
b) $\frac{1}{5}$
c) $\frac{n}{n - 3}$
d) $\frac{x - 4}{x + 3}$

e) $\frac{2x + 3}{x + 1}$
2 a) $\frac{8c + 3b}{4bc}$
b) $\frac{9 - y^2}{3y}$
c) $\frac{7x + 3}{10}$

d) $\frac{n + 4}{(n - 5)(n - 2)}$
e) $\frac{5x + 2}{2x^2}$
f) $\frac{7}{2d}$
g) $\frac{3u}{10v}$

Chapter 5

1 a) $x = -7$
  b) $x = -\frac{1}{2}$
  c) $x = 22$
  d) $x = 11$
  e) $x = \frac{-10}{3}$
  f) $x = 2$

2 a) $x \leq 4$
  b) $x > \frac{-3}{4}$
3 $m = \frac{k + 3n}{4}$
4 $B = \frac{P - 2L}{2}$
5 $h = \frac{3V}{A}$
6 $x = b(a + c)$

7 $v = \sqrt{\left(\frac{T - 2}{u}\right)}$
8 $n = \sqrt{\left(\frac{5T}{k}\right)}$
9 $q = (L - p)^2$

Chapter 6

1 $\frac{3}{4}$
2 a) $y = 4x - 2$
  b) $y = -\frac{2}{3}x + 4$
  c) $L = \frac{1}{2}n + 2$
  d) $y = 2$

  e) $x = -1$
  f) $f(x) = 7 - 2x$
3 a) $y = 5x + 13$
  b) $x = 4$

4 a) $m = \frac{7}{4}$, y-intercept is $(0, 2)$
  b) $m = \frac{-1}{2}$, y-intercept is $(0, 5)$
5 $b = 7, a = 8$

Chapter 7

1 $x = 4, y = -2$
2 $x = 2, y = 1$

3 a) $x = 3, y = -1$
  b) $x = 5, y = 2$
4 a) $3x + 2y = 26$, $4x + 7y = 52$
  b) adult $= £6$, child $= £4$

5 a) $5c - 2i = 51, c + i = 20$
  b) 13 questions correct,
      7 questions incorrect
6 $(8, 3)$

Chapter 8

1 $y = 5x^2$
2 $y = -3x^2$
3 a) $y = 3(x + 2)^2 - 6$
  b) $y = 3 - (x - 1)^2$
4 a)

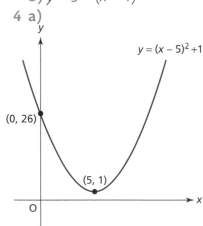

b)

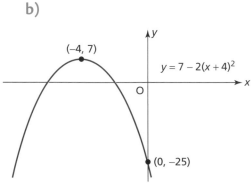

5 $A(-1, 0)$, $B(7, 0)$, $C(0, -7)$,
    turning point $(3, -16)$

6 a)

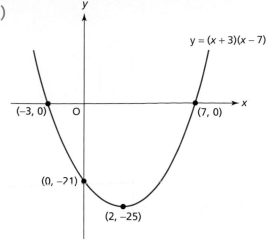

b)

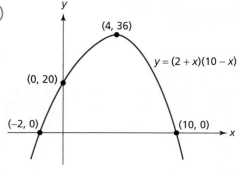

7 $x = \frac{1}{2}$

Chapter 9

1 a) $x = -3, x = 2$
   b) $x = -1, x = 4$
2 a) $x = 0, x = 8$
   b) $x = \frac{-3}{2}, x = \frac{3}{2}$
   c) $x = 6$
   d) $x = -4, x = 7$

   e) $x = \frac{-2}{3}, x = -1$
3 a) $x = 0.4, x = -1.2$
   b) $x = 4.3, x = 0.7$
4 a) two real and distinct
   irrational roots

   b) no real roots
5 two equal real roots
6 two real and distinct rational
   roots
7 10 cm
8 $(-1, 2), (3, 14)$

Chapter 10

1 a) $x = 6.7$
   b) $x = 8.5$
2 UV = 9.2
3 diameter = 30 cm
4 CI = 13.9 cm

5 a) is not right-angled (*note:*
   $9^2 + 15^2 = 306$; $17^2 = 289$)
   b) is right-angled (*note:*
   $2.8^2 + 4.5^2 = 28.09$; 5.3^2
   $= 28.09$)

6 ABCD is a rectangle (*note:*
   $4.5^2 + 6^2 = 56.25$; $7.5^2 = 56.25$)

Chapter 11

1 75°
2 50°

3 120°
4 110°

5 interior angle = 135°, exterior
   angle = 45°

Chapter 12

1 28.8 cm
2 32.1 m^2
3 70°
4 140°

5 99.0 cm
6 17.0 m^2
7 53°
8 136°

9 12.5 cm
10 5 m

Chapter 13

1 22.5 cm
2 12 cm

3 10.8 m^2
4 1200 cm^3

5 £2.40
6 £1.00

Chapter 14

1 268 m^3
2 109.2 m^3

3 436 litres
4 0.838 m^3

5 7.98 cm
6 29

Chapter 15

1 a) $\begin{pmatrix} 5 \\ -2 \end{pmatrix}$

 b) $\begin{pmatrix} -2 \\ 4 \end{pmatrix}$

 c) $\begin{pmatrix} 3 \\ 0 \end{pmatrix}$

2 a) $\overrightarrow{PS}$
 b) $\overrightarrow{UQ}$
 c) $\overrightarrow{TP}$

3 a) c
 b) d
 c) −b

4 a)

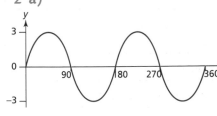

 4 km/h
 7·5 km/h
 North

 b) speed = 8·5 km/h,
 bearing = 073°

5 A(4, 0, 0), B(4, 2, 0), C(0, 2, 0),
 D(2, 1, 7)

6 a) $\begin{pmatrix} 2 \\ 3 \\ 0 \end{pmatrix}$

 b) $\begin{pmatrix} 2 \\ 0 \\ 5 \end{pmatrix}$

 c) $\begin{pmatrix} 0 \\ -3 \\ -5 \end{pmatrix}$

 d) $\begin{pmatrix} 2 \\ 3 \\ -5 \end{pmatrix}$

 e) $\begin{pmatrix} -2 \\ 3 \\ 5 \end{pmatrix}$

7 a) $\begin{pmatrix} -1 \\ 1 \\ 9 \end{pmatrix}$

 b) $\begin{pmatrix} -3 \\ 5 \\ 1 \end{pmatrix}$

 c) $\begin{pmatrix} -1 \\ 0 \\ 22 \end{pmatrix}$

8 a) $4\sqrt{5}$
 b) 7

Chapter 16

1 a) 966 m²
 b) 55·4 m
2 35·7 km
3 33°

4 70·8°
5 69·6°
6 192 km

7 distance = 575 km, bearing =
 298°
8 18·5 m

Chapter 17

1 a) $y = 2\cos 4x°$
 b) $y = 5\sin x° - 2$
 c) $y = \tan 2x°$ or $y = a\tan 2x°$
 d) $y = 4\cos(x + 60)°$

2 a)

b)

3 A(90,12), B(270,2)
4 a) 14·5°, 165·5°
 b) 158·2°, 338·2°
 c) 128·7°, 231·3°

5 p = −0·5, q = 250·5
6 a) $\frac{5}{13}$
 b) $\frac{12}{5}$
7 $\tan x°$
8 LHS = $\cos x° \sin x° + \cos^2 x°$
 $+ \sin^2 x° - \sin x° \cos x° =$
 $\cos^2 x° + \sin^2 x° = 1$ = RHS

Chapter 18

1 a) median = 131 cm,
 SIQR = 8.5 cm
 b) The girls are taller (on
 average) and their heights
 vary more than the boys'
 heights.

2 a) mean = 71, standard
 deviation = 1·7 (1 d.p.)
 b) Jack's scores are lower
 (on average) and they are
 more consistent.

3 a) $y = -1750x + 18500$
 b) £2750

Overview of National 4 Mathematics Course

The Course comprises three component Units – Expressions and formulae, Relationships and Numeracy. A fourth Unit, the Added Value Unit, is assessed by means of a test. To gain the Course award, you must pass all four of the Units. The Units are assessed internally on a pass/fail basis within centres. National 4 courses are not graded.

The following tables give details of the skills, knowledge and understanding covered in the National 4 Mathematics Course.

National 4: Expressions and formulae Unit	Explanation of Standard
Algebraic skills	
Expand brackets with a numerical common factor	*e.g. 3(4x + 2), 5(a − 2c)*
Factorise a sum of terms with a numerical common factor	*e.g. 7x + 21, 24y − 9*
Simplify an expression which has more than one variable	*e.g. 3a + 4b− a + 6b*
Evaluate an expression or a formula which has more than one variable	*e.g. 4w + 6l − 3k*
Extend a straightforward number or diagrammatic pattern and determine its formula	*Sequences such as 4, 7, 10, 13, ...* *Evaluate the formula for a given value.*
Calculate the gradient of a straight line from horizontal and vertical distances	*Vertical distance over horizontal distance*
Geometric skills	
Calculate the circumference and area of a circle	
Calculate the area of a parallelogram, kite, trapezium	
Investigate the surface of a prism	*Face, vertex, edge; draw nets; calculate surface area*
Calculate the volume of a prism	*Triangular prism, cylinder, other prisms given the area of the base*
Use rotational symmetry	
Statistical skills	
Construct a frequency table with class intervals from raw data	*Using ungrouped data*
Determine statistics of a data set	*Mean, median, mode, range*
Interpret calculated statistics	*Use mean, median, mode, range to compare data sets.*
Represent raw data in a pie chart	
Use probability	*Calculate probability and interpret it in the context of risk.*
Reasoning skills	
Interpret a situation where maths can be used and identify a valid strategy	*Attach to any of the above skills.*
Explain a solution and/or relate it to context	*Attach to any of the above skills.*

National 4: Relationships Unit	Explanation of Standard
Algebraic skills	
Draw and recognise a graph of a linear equation	Draw a graph of values for chosen values of x. For $y = mx + c$, know the meaning of m and c. Recognise and use $y = a$, $x = b$.
Solve linear equations	$ax + b = c$, $ax + b = cx + d$ where a, b, c, d are integers
Change the subject of a formula	e.g. $G = x + a$ to x, $h = \frac{v}{n}$ to v, $E = 3w - k$ to w
Geometric skills	
Use Pythagoras' theorem	Given measurements or coordinates
Use a fractional scale factor to enlarge or reduce a shape	Non-regular rectilinear shapes
Use parallel lines, symmetry and circle properties to calculate angles	Combination of angle properties associated with: • intersecting and parallel lines • triangles and quadrilaterals. Circles: • angles in a semi-circle • relationship between tangent and radius.
Trigonometric skills	
Calculate a side in a right-angled triangle	Given a side and an angle.
Calculate an angle in a right-angled triangle	Given two sides.
Statistical skills	
Construct a scattergraph	Given a set of data.
Draw and apply a best-fitting straight line	Use the line of best fit to estimate one variable, given the other.
Reasoning skills	
Interpret a situation where maths can be used and identify a valid strategy	Attach to any of the above skills.
Explain a solution and/or relate it to context	Attach to any of the above skills.

National 4: Numeracy Unit	Explanation of Standard
Use the following skills to solve straightforward real-life problems involving money/time/measurement.	
Numerical outcome	
Select and use appropriate numerical notation and units	• numerical notation should include: $=$, $+$, $-$, $\times$, $\div$, $/$ $<$, $>$, $( )$, %, decimal point • units should include: • money (pounds and pence) • time (months, weeks, days, hours, minutes, seconds) • measurement of length (millimetre, centimetre, metre, kilometre, mile) • weight (gram, kilogram), volume (millilitre, litre) and temperature (Celsius or Fahrenheit)

National 4: Numeracy Unit	Explanation of Standard
Select and carry out calculations	<ul><li>add and subtract whole numbers including negative numbers</li><li>multiply whole numbers of any size, with up to four-digit whole numbers</li><li>divide whole numbers of any size, by a single digit whole number or by 10 or 100</li><li>round answers to the nearest significant figure or two decimal places</li><li>find simple percentages and fractions of shapes and quantities e.g. 50%, 10%, 20%, 25%, $33\frac{1}{3}$%, $\frac{1}{2}$, $\frac{1}{3}$, $\frac{1}{4}$, $\frac{1}{10}$, $\frac{1}{5}$</li><li>calculate percentage increase and decrease</li><li>convert equivalences between common fractions, decimals and percentages</li><li>calculate rate e.g. miles per hour, texts per month</li><li>calculate distance given speed and time</li><li>calculate time intervals using the 12- and 24-hour clock</li><li>calculate volume (cube and cuboid), area (rectangle and square) and perimeter (shapes with straight edges)</li><li>calculate ratio and direct proportion</li></ul>
Read measurements using a straightforward scale on an instrument	<ul><li>measure length, weight, volume and temperature</li><li>read scales to the nearest marked, unnumbered division with a functional degree of accuracy</li></ul>
Interpret measurements and results of calculations to make decisions	<ul><li>use appropriate checking methods e.g. check sums and estimation</li><li>interpret results of measurements involving time, length, weight, volume and temperature</li><li>recognise the inter-relationship between units in the same family, e.g. mm/cm, cm/m, g/kg and ml/l</li><li>use vocabulary associated with measurements to make comparisons for length, weight, volume and temperature</li></ul>
Explain decisions based on the results of measurements and calculations	give reasons for decisions based on the results of calculations
Graphical data outcome	
Extract and interpret data from straightforward graphical forms	including tables (with at least four categories of information), charts (e.g. pie), graphs (e.g. bar, line, scatter) and diagrams (e.g. stem and leaf, map, plan)
Make and explain decisions based on the interpretation of data from straightforward graphical forms	<ul><li>make decisions based on:<ul><li>observations of patterns and trends in data</li><li>calculations involving data</li><li>reading scales in straightforward graphical forms</li></ul></li><li>offer reasons for the decisions made based on the interpretation of data</li></ul>
Make and explain decisions based on probability	<ul><li>recognise patterns and trends and use these to state the probability of an event happening</li><li>make predictions and use these to make decisions</li></ul>

National 4: Value Added Unit	Content/Skill
Operational skills	
Numeracy and Statistics (non-calculator)	• *round solutions to a given number of decimal places* • *calculate whole number percentages of quantities (single digit percentages, 50%, multiples of 10% and 25%)* • *calculate the mean of a data set requiring rounding* • *calculate simple fractions of quantities* • *add two decimal numbers (with a differing number of decimal places) and then subtract from the result* • *multiply a decimal number by a single-digit whole number*
Algebra, Geometry and Trigonometry	• *linear equations which require the use of distributive law and simplification* • *problems using area (circle, parallelogram, kite, trapezium), surface area (prism) or volume (prism)* • *create and use a formula (from a numerical or diagrammatic sequence)* • *calculations involving speed, distance and time, where time is given or calculated as hours and minutes* • *use Pythagoras' theorem in problems* • *use trigonometry to calculate a side or angle of a right-angled triangle* • *solve problems involving the completion of a described shape using coordinates spread over at least three quadrants*
Reasoning skill	
Identify a valid strategy or explain a solution	• *attach appropriately to questions involving operational skills*